PHYSICAL GEOGRAPHY AND
CLIMATOLOGY

The Aletsch Glacier, Switzerland, with the Jungfrau, Mönch and Trugberg in the
background and the Marjelensee in the foreground

PHYSICAL GEOGRAPHY AND CLIMATOLOGY

by

N. K. HORROCKS, B.Sc.

*Senior Geography Master, Central Grammar School,
Manchester*

With a Foreword by
PROFESSOR S. W. WOOLDRIDGE, D.Sc.

LONGMANS, GREEN AND CO.

LONDON   NEW YORK   TORONTO

LONGMANS, GREEN AND CO LTD
6 & 7 CLIFFORD STREET LONDON W 1

ALSO AT MELBOURNE AND CAPE TOWN

LONGMANS, GREEN AND CO INC
55 FIFTH AVENUE NEW YORK 3

LONGMANS GREEN AND CO
215 VICTORIA STREET TORONTO 1

ORIENT LONGMANS LTD
BOMBAY CALCUTTA MADRAS

First published 1953

PRINTED AND BOUND IN GREAT BRITAIN
BY HAZELL WATSON AND VINEY LTD
AYLESBURY AND LONDON

To
B. B.
in
friendship

FOREWORD

by

PROFESSOR S. W. WOOLDRIDGE, D.SC.

I am glad to write an introduction to Mr. Horrocks's book because I hope that it will play its part in a revival of the teaching of Physical Geography in schools. Such a revival is now overdue. For many years we have followed a principle established by the pioneers of geographical teaching in Britain that Physical Geography should be subordinate to Regional Geography. Many like myself, proud to number themselves as among the students of James Fairgreave, will remember his insistence that the 'capes and bays geography' of old unhappy times bade fair at one stage to give place to a rather cold and abstract regimen of 'scientific capes and bays', isotherms, isobars, *et hoc genus omnes*. He and his associates were quite right in their emphasis, but we who learned from them have tended to take a good principle and carry it to extremes. It is one thing, and often a good thing, to teach Physical Geography incidentally, as it were, to Regional Geography; it is quite another to ignore it altogether. Those of us who teach geography in the Universities and are concerned for the good name and future progress of our subject have all too often been confronted with classes essentially ignorant of the build of their planet, the significance of its landscapes, and the mechanism of its climates. These have been ignored as topics too abstract, difficult, or even irrelevant to find place in an admittedly crowded course and University Departments have necessarily had to teach their Physical Geography from the beginning.

I am far from suggesting that the sole purpose of school geography is to provide entrants for the Honours Schools of the Universities. The case for Physical Geography rests on much better and wider grounds than this. It is of high cultural value and upon it rests most of the practical applications of geography of which the modern world finds itself in need. We may perhaps recall G. K. Chesterton's mildly ironical characterization of the geography he knew:

"Our principal imports come far as Cape Horn,
For necessities cocoa, for luxuries corn.

Thus Brahmins are born for the ricefield and thus
The Gods made the Greeks to grow currants for us."

Even in these wider fields we shall find that as geographers we can
have little very significant to say about Brahmins, Greeks, or currants
without the underlying discipline and control of Physical Geography.

Mr. Horrocks's book is systematic and detailed. He presents his
matter in the light of his experience as a teacher of many years'
standing. It is natural and even proper that his friends and colleagues
will not always agree with the matter or treatment which commends
itself to him. None the less I find myself in substantial agreement
with the method and manner he has used and, what is far more
important, I am assured that no intelligent boy or girl at the VIth
Form stage could read this book without a growing realization that
our earth is something more than the sum of its parts, and that
geography offers a field of study very attractive and amply repaying
to the intelligent and enquiring mind.

I hope this book will be widely used, for I know that if it is, the
objectives of geographical education will be soundly advanced.

S. W. W.

PREFACE

The purpose of this book is to provide an account of the several aspects of physical geography which shall be as comprehensive as possible. It is intended primarily for pupils in secondary grammar schools who are preparing for an examination in geography at an advanced level. Nevertheless, it is hoped that some parts of the text may be of value to those pupils who will reach scholarship standard.

In order to give breadth to the discussion of most of the topics it has been necessary to make some selection of material. Thus no systematic account is given of mathematical geography (it is assumed that the reader is in possession of the elementary facts) so that where necessary, as in Chapter X, only the essential facts are stated, quite briefly, in the text. Also, it has not been easy, in Chapter I, to limit the discussion of the basic facts of geology, but the choice which has been made should be sufficient for reading the following chapters in Parts I and II of the book. Rather more detailed treatment has been given to several matters of basic importance which experience has shown can present difficulties to most senior pupils. Throughout the writing of this book every endeavour has been made to present the material in such a manner as to make it possible for the reader who is not studying science at an advanced level to gain an adequate knowledge of physical geography. Every care has been taken to establish a logical sequence. However, the accounts of both vegetation and soils in Part IV are incomplete without reference to related sections in Part V. Furthermore, in order to avoid an unduly large number of cross-references in the text, a limited amount of repetition has been made. It has been decided to omit world distribution maps of temperature, pressure, etc., as these are available in advanced atlases.

The author is indebted to many friends, colleagues, and others for their invaluable assistance in one way or another. The list of these would, indeed, be long if given in full, but a particular debt of gratitude is owing to the following: Professor E. Ashby (who gave much advice for Chapter XV); and to Messrs. E. P. Boon, L. Hadlow, G. C. Rosser, W. Shercliff, M. L. T. Sinclair, A. W. Smith, H. Smith, and R. P. Trueblood. Thanks must also be recorded to Professor Wooldridge for his guidance and his interest in this work of one of his past students; to Professor A. A. Miller, for permission to draw freely on his *Climatology* for statistics and other material; and to Professor S. H. Beaver. Needless to say the author is responsible for the form in which the material is presented.

The diagrams and pictures have been chosen so as to give as fresh an aspect as possible to the book. However, in several instances it has been considered wise to make use of existing illustrations from other books. Recognition of these and of the sources of photographs is made where necessary and the author thanks all concerned. In some cases sources cannot be traced and it is regretted that acknowledgment is therefore not possible.

N. K. HORROCKS.

Sale, Cheshire.
May 1st, 1953.

CONTENTS

Part One

STRUCTURE OF THE EARTH'S SURFACE

Part Two

EARTH SCULPTURE

Part Three

METEOROLOGY

Part Four

PLANTS AND SOILS

CONTENTS

Part Five

CLIMATOLOGY

Part Six

OCEANOGRAPHY

Part I

STRUCTURE OF THE EARTH'S SURFACE

CHAPTER I

INTRODUCTION—WEATHERING PROCESSES AND ROCKS

Introduction

In approaching the study of physical geography, it is desirable to make a brief reference to some of the theories and speculations concerning the origin of the earth and kindred topics. These problems rightly belong to the studies of astronomy and cosmogony, and such theories as those of Kant, Laplace (Nebular Hypothesis), and the Planetesimal Hypothesis cannot justifiably be discussed at length in this book.

It is widely accepted that the earth was originally in the form of a hot, gaseous mass which cooled to a liquid in a relatively short time as measured on the time scales of astronomy and geology (whose units are millions of years rather than centuries).

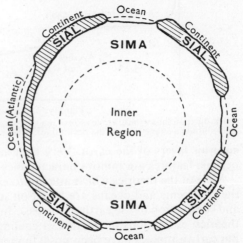

FIG. 1.—The Zones of the Earth (not to scale)

Later, a solid crust was formed, on which an atmosphere with water vapour collected, thus providing surface water by condensation.

Our knowledge of the interior of the earth is very limited. The products of volcanic eruptions help to give some idea of the probable chemical and physical conditions below the surface, and the investigations of seismologists have led to the belief that the interior has concentric layers of material; the density of the consecutive layers increases with depth. Observations by geologists suggest that there is a surface layer, represented by the continental masses, which is distinct from the layer immediately below; the upper surface of the second layer forms the floors of most oceans (Fig. 1).

The continental material is often called sial and that of the ocean floors is known as sima. These two terms are derived from the names of the elements which are most common in each (viz. *si*licon and *al*uminium, and *si*licon and *ma*gnesium respectively) as determined by the chemical analysis of the rock material.

By estimation of the areas of land and ocean, within certain ranges of height and depth respectively, it is possible to plot a hypsographic curve (Fig. 2) which gives a generalized cross-section of the continents and the ocean basins. It will be observed that there are two quite definite levels coinciding with the continental surfaces and the ocean floors which gives support to the suggestion of the two

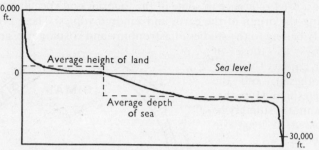

FIG. 2.—The Hypsographic Curve

The diagram shows the average level of the land surface at about 2,750 feet above sea-level and the average level of the ocean floor at about 12,500 feet below sea-level.

superior layers of the earth. Very high land features and very low oceanic features are minor characteristics of this curve.

Most of the theories which attempt to explain the arrangement of the continents and oceans are based on some aspect of the cooling and contraction of the earth in the past. In the development of these theories, estimations of the rate of cooling of the earth were made, but early estimates have been upset in more recent times by the new knowledge of a source of heat within the earth which was previously unsuspected. It is now known that scattered amongst the materials of the various layers of the earth are radioactive elements, which are, and have been, liberating heat spontaneously as a result of atomic disintegration. The most effective elements in this respect are uranium and thorium. But potassium, although less effective, is far more widespread in the rocks than either of the others, and therefore probably makes a considerable contribution to this heat. This spontaneous disintegration is a constant process, and forms a time measure for the geologist, who has now extended the age of the oldest rocks to some two thousand million years.

The discovery of this internal supply of heat has modified our views of the nature of the material in the deeper zones of the earth. This material must remain at very high temperatures, since there is this constant supply of heat to make good that lost by conduction to the surface. The materials below the upper layers must be so hot that they are capable of flow under great pressure, although it must be appreciated that at these pressures the melting-points of these materials will be raised, and so higher temperatures will be required than those which would produce melting at the surface of the earth. The materials are not liquid as commonly understood, but are capable of movement under great stresses. This conception of the plasticity of the sima in contrast to the rigidity of the sial may help to explain

the separation of the earth's surface into continental masses and ocean basins. A glance at Fig. 1 will show that the sial is light and that it rests on the sima, which is more dense. We may regard the masses of sial as rafts 'floating' in the sima, which, by virtue of its inherent radioactive heat, can become at times almost 'liquid'.

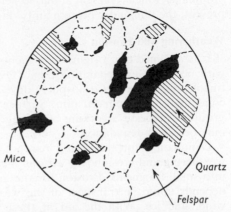

FIG. 3.—A Coarse-grained Igneous Rock as shown by a Microscope Slide

The possibility of lateral movement of a sial raft in the sima sub-stratum has been postulated by Wegener in his theory of continental drift. Wegener found it necessary to envisage a dispersal of the continents, as we see them today on a world map, from a single great land mass of the past in order to explain various apparent discrepancies arising from his study of fossil plants. The mechanism by which this movement of the continents has been effected is not known, but many suggestions, apart from those of Wegener, have been put forward.

Minerals in Rocks

The sial rafts may be regarded as masses of a coarse-grained igneous rock, formed by the cooling of molten material. Examination of such rocks shows that they are made up of various minerals in the form of crystals, which differ in size and chemical composition. Some are large, whilst others are relatively small; some are dark in colour, whilst others are light (Fig. 3). In specimens of what, for the moment, may be called sialic rock, light-coloured minerals are generally predominant. Of these, two are very common: (*a*) quartz, which is largely silica, or the oxide of the element silicon, and (*b*) felspar, which contains alumino-silicates, or compounds of silicon with aluminium and oxygen, in association with calcium, potassium, and sodium in varying proportions. Thus there may be several types of felspar in any one specimen of rock. The dark colours are indicative of minerals, such as mica, which are also rich in silicon and aluminium, but dark by virtue of iron oxides and other compounds. A common rock of this type is granite, which has approximately equal proportions of quartz and felspar (the latter is slightly more predominant), which represent about 80 per cent. of the various minerals present, the remainder being mainly dark in colour.

Weathering and Erosion

It is assumed that original continents of sial came into existence with ocean basins separating them. Over the surface of both continents and oceans an atmosphere developed in which water vapour condensed and produced rain. Surface changes were effected by a group of processes (working both singly and in partnership) which may be called denudation.

Denudation (or the laying bare of the surface) involves both weathering and erosion. By various means the rocks are weathered into small fragments. Erosion (or the gnawing away of the surface) depends upon the transport of the weathered fragments by water, wind, and ice. With the aid of these fragments the transporting agents can erode or reduce other parts of the earth's surface.

Mechanical Weathering

The all-important influence of weathering may be due to purely mechanical processes such as:

1. The disintegration of the surface of rocks caused by their exposure to the intense heating by the sun. The minerals in rocks expand by varying amounts so that great forces are created, and some of the particles are ejected from the rock surface. Although this process is universal, it is most common in the hot deserts of the world, where sometimes the rock peels off in layers. This peeling effect is known as exfoliation.

2. At night the opposite effect is frequently produced as a result of intense chilling, and then mineral particles are loosened by the forces developed by excessive contraction.

3. Frequently this last process is assisted by frost action. In surface crevices water collects and with nocturnal chilling, if this is suffi-

ciently great, a wedge of ice is formed. Since the water expands on freezing, very considerable forces are brought to bear on the mineral particles on the sides of this wedge of ice, and it is usual for these forces to loosen some of the particles, although they are not free to be moved until the ice melts with a rise of temperature. Large-scale frost action is widespread on high mountains, and can result in the formation of extensive layers of shattered rock fragments known as screes. Small-scale frost weathering is almost world-wide, and in temperate and cold lands it may be of vital assistance to the farmer in breaking up the soil.

Chemical Weathering

The chemical nature of the various minerals in a rock determines the extent to which weathering may be produced by the action of rain-water. Rain is not merely water, but must be regarded as a mildly acid solution.The following chemical changes may occur:

1. Some mineral particles are removed in solution from the rock.

2. Other minerals absorb water, and new compounds are formed by the process of hydration. Many of these freshly formed substances are soluble in rain-water, whereas before hydration the minerals from which they were formed were insoluble. After hydration some minerals show a great increase in bulk, and are more readily attacked by the agents of mechanical weathering.

3. Since rain-water contains oxygen, there is a considerable oxidation of some minerals and many oxides are produced, particularly those of iron, which are red, yellow, and brown in colour.

4. Carbon-dioxide is another gas which is dissolved in rain-water. The weak acid solution, which is thus formed, produces carbonation, and amongst other compounds, carbonates and bicarbonates are developed. Most of these products are then easily dissolved in the rain-water. Carbonation is of major importance in the weathering of limestone rock (see p. 127).

The result of mechanical and chemical weathering is to convert and rearrange the minerals of the original rock into particles of various compounds. Some of these particles are produced directly by either mechanical or chemical action. Others are weathered first to a rock waste of various minerals[1] by mechanical action and then further changed by chemical action. As will be seen later the end-products may be regarded as particles of clay,[2] sand,[3] and carbonates. Common rocks, such as sandstone and grit, are developed

[1] Mainly compounds of silicon, aluminium, calcium, iron, and oxygen.
[2] Hydrated aluminium silicate.   [3] Silica with oxides of iron.

from sand fragments; shale from clay fragments; and chalk and limestone from the carbonates.

Some Forms of Erosion

It is necessary to pause for a moment and consider other mechanical processes which assist in the complete weathering of a rock surface. Without the preliminary comminution of rock particles that has been described, these latter processes could hardly proceed.

Wind is an agent of erosion, because when small hard fragments of rock waste are blown against other rock surfaces there is abrasive action which will reduce the less-resistant particles in that rock. Where soft sandstone and brickwork are employed for the facings of buildings, the scouring effect of wind-driven particles may be very apparent. Wind plays its part in some localities on a grand scale, as in deserts and in some coastal districts (see pp. 119 and 140).

Rain-drops, when driven by the wind, behave in very much the same fashion, and drops falling with very heavy rainfall may pro-

FIG. 4.—Formation of Earth Pillars

duce a form of erosion which is quite local, giving rise to the spectacular earth pillars (Fig. 4). Such pillars are found in the Italian Alps near Bolzano, and some small ones are known in the Findhorn valley of Scotland.

Rock waste is eventually transported to a stream by rain-wash, either as insoluble solid fragments in suspension or as dissolved fragments in solution. The fragments in suspension are of various sizes; some are so massive that they move only when the river flows fast, and even then their movement is more akin to rolling and sliding along the river-bed. This movement of solid fragments is important, in that it leads to further wearing down as the fragments are projected on to rock surfaces in the river course. This is often referred to as corrasion.

The work of a river is of such importance that a detailed study is made of it in a later chapter, but for the moment we must follow the suspended fragments as the river makes its way to the sea. In general, a river slows down as it approaches the sea, and this means that it can only carry quite small fragments in suspension and, of course, those solids which are in solution. On entering the sea, the flow of the river is almost completely halted. The coarser fragments in suspension are jettisoned as sand, and the very finest move out into the sea for considerable distances before sinking to the sea floor to form muds and clays.

Ice is capable of transporting rock fragments and even large rock masses. By gripping and enclosing fragments of all sizes within its mass, a sheet of ice can erode extensively any rock surface over which it moves. The scope of such ice erosion is limited by temperature and ceases with the melting of the ice. Today the areas of the world which are so affected are restricted to high latitudes or high altitudes.

Sedimentary Rocks

It is on the sea floor that new rock formations are produced from the assorted particles derived from the land. In addition, the dissolved solids which have been carried from the land into the sea may be extracted from solution by sea organisms in order that they may build their shells and other body structures of calcium carbonate or silica. When these organisms die, their shells and hard structures are laid on the sea floor, thus forming layers of carbonates, although some portions of these layers are formed by the physical condition whereby dissolved carbonates are deposited directly from sea water with an increase of temperature.[1] Some solids remain in solution, thus increasing the salinity of the ocean waters.

Thus are formed new rocks of the type known as sedimentary. Grains of sand which are so accumulated become sandstone and grit by a process of cementation effected largely by iron oxides. Muds become clays and shales as a result of compression of one layer upon another. Carbonates produce limestone and chalk rock.

Marine Transgressions

At this point it may be asked how such rocks are found today on the land surfaces. The continents may suffer a relative uplift, due to forces which have their application in or below the sialic rafts. Such an uplift of an original continent would expose inshore deposits that had already accumulated as waste from the land in the marginal seas. On the other hand, a depression of a sial block would permit its marginal areas to be flooded by the sea. Such flooding is known as a marine transgression, and in these marginal seas would be laid fresh deposits from the land. Renewed uplift would expose land areas

[1] It is probable that the presence of ammonia assists this deposition.

with a cover of newer surface rocks (Fig. 5). Such a sequence of events must have occurred on many occasions in the past. The process of marine transgression is dealt with more fully on p. 21.

Thus the continents have acquired, by a series of such transgressions, a great number of layers of sedimentary rock. These in their turn have undergone weathering and erosion as described above, so that still newer layers can be built up in the course of long time.

Not all sedimentary rocks are laid in marginal seas; some have been formed entirely on the land surfaces, such as certain sandstones

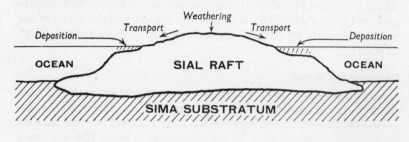

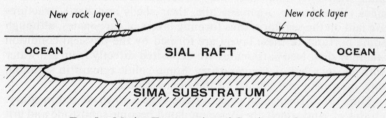

FIG. 5.—Marine Transgression of Continent (above)
Retreat of Sea after Transgression (below)

which reflect past desert climates. Others indicate that they have been formed from the coarse material laid down by a previous river in its delta.

Rock Structures

It is evident that the sedimentary rocks would be laid in approximately horizontal layers or strata. These rocks do show stratification, but on land they may be inclined at considerable angles to the horizontal, possessing, in geological parlance, a dip (Fig. 6). In some rock exposures it is possible to note that the dip changes in direction (Fig. 7), and in very favourable sections one may note that the strata may exhibit the appearance shown in Fig. 8. The latter example shows folding of the rocks, and this is due to the compressional forces which must act, presumably, parallel to the surface of the earth. In other rock exposures it is possible to see that the strata in two adjacent sections are displaced in relation to one another (Fig. 9). This is an example of faulting, and is the result of tensional or

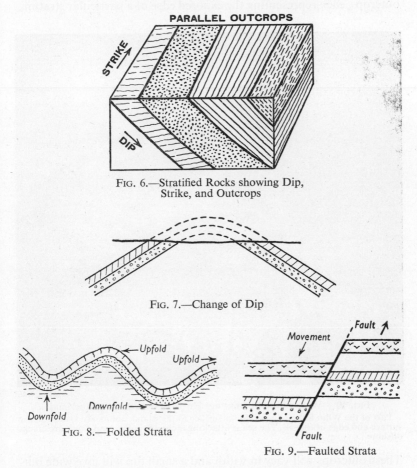

FIG. 6.—Stratified Rocks showing Dip, Strike, and Outcrops

FIG. 7.—Change of Dip

FIG. 8.—Folded Strata

FIG. 9.—Faulted Strata

compressional forces acting vertically, obliquely, or horizontally, to the surface. The cracks or faults which are developed may penetrate deeply into the sial mass. Similar forces may produce jointing, that is to say, the development of intersecting vertical cracks in the rocks (Fig. 10). Limestone is an example of a well-jointed rock, and the joints divide it up into a series of massive blocks. These joints are of great assistance in the chemical weathering of limestone, since they give access to percolating rain-water. In some instances joints may form with the shrinkage of rocks as they dry out, as cracks form in dried mud.

Where rocks have a dip, erosion may reveal a surface with several outcrops, each representing the exposed edge of a particular stratum.

Aerofilms, Ltd.

FIG. 10.—Whin Sill with Jointing, Farne Islands, Northumberland

Part of the Whin Sill giving a tabular surface to the island. Joints are visible on the surface and edge of the sill. The sea is attacking along the joints and there are stacks offshore.

These outcrops will vary in width and a small dip will give wide out-crops, whereas if the same strata are dipping at a high angle, the outcrops will be relatively narrow. The surface trend of an outcrop is known as the strike of the rock (Fig. 6).

Igneous Rocks

Faulting which has penetrated the sial blocks may give rise to two other features deserving of comment. These cracks give access to the surface of molten material from the hot inner layers of the earth. Thus, in certain localities, outpourings of lava or the building of volcanoes may occur. The molten material is sometimes sidetracked on its way to the surface and seeps in between two strata (Fig. 16), producing a sill of rock. The material which rises up a fault may solidify without giving a lava flow on the surface. It then produces a narrow wedge or dyke, which when exposed cuts the surface as a thin outcrop. In either case, whether the surface is reached or not, the molten material cools and solidifies to form a rock which is like granite, in that it is crystalline. To the geographer it is sufficient to distinguish this rock as lava or basalt. Basalt differs from granite in having mineral particles which are minute and essentially microscopic. It has also a dark greenish-black or greenish-blue colour. Quartz is less common than it is in granite, but felspar is very important. In addition there are various dark-coloured minerals formed from compounds of iron and magnesium. This assemblage of minerals may be regarded as typical of the sima which, it will be remembered, lies under the sialic rafts and forms the floors of the ocean basins.

Both granite and basalt are known as igneous rocks, and may be regarded as a mass of crystals which have formed from a hot mixture of minerals (molten rock) that has cooled. Various factors, such as the rate of cooling, the pressure, and the original composition of the molten rock, have resulted in the large crystals of granite, with its preponderant quartz and felspars, and in the small crystals of the basalt with its preponderant felspars and dark minerals.

Metamorphic Rocks

Sedimentary and igneous rocks may be altered or metamorphosed by a variety of processes, details of which are largely of geological importance. Metamorphism means, literally, a change of form. It shows itself in rocks in various ways:

1. By foliation, whereby the rock has a distinct grain picked by the alignment of mineral particles, usually following the stratification, but often in wavy lines.

2. By cleavage, which resembles foliation, although the direction of cleavage may be quite independent of the stratification.

3. By the development of new materials which were not present in the unaltered rock; precious stones and valuable ores may be produced by metamorphism.

4. By the development of a crystalline structure from an originally amorphous rock.

5. By the development of amorphous material from an originally crystalline rock.

Sometimes more than one process may be responsible for the metamorphism. Intense heat (developed by the contact of molten rock), intense strain (from severe earth forces), solutions charged with gases (and even very hot water) may be involved, singly and severally. The physical effect of heat and strain, combined usually with chemical reactions produced by solutions, appear to result in a very profound change in the chemical or molecular structure of the various minerals within the rocks undergoing metamorphosis, so that in some cases one may say that essentially new rocks are formed.

Almost every type of sedimentary rock has a metamorphosed form. Limestone and chalk may be converted into marble. In Weardale, Co. Durham, the baking of limestone by the injection of basalt in the Whin Sill has caused the rock to become crystalline.

Shales may be converted to slate, which is a more valuable rock, since its cleavage makes it easy to split into uniformly thick slabs (Fig. 11). Sandstones and grits may develop a coarse crystalline texture when they are changed into quartzite. Coal is possibly converted into anthracite by metamorphic processes. Igneous rocks may yield veins of valuable ores through metamorphosis, and steam charged with acids may convert crystalline granite into amorphous kaolin or china clay.

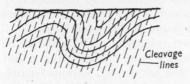

FIG. 11.—Folded Slates with Cleavage

These particular changes are fairly localized, although extensive areas of slate are recognized in Wales and elsewhere. Over a large area, such as the Highlands of Scotland, widespread metamorphism has given rocks which are distinguished by the geologist, but which may be regarded by the geographer as granitic, in that the essential features of these rocks are those of granite. Vast areas of the Canadian Shield, the Brazilian and the Guiana Plateaus have suffered from this regional metamorphism, and consist mainly of granitic rocks with local areas of particular value. Such areas are at Sudbury with its nickel ores and at Cobalt with its silver and cobalt deposits in Ontario; at Itabira, in Brazil, there are iron ores, whilst in similar regions there are to be found the gold-mining areas of the world.

Stratigraphical Sequence

One branch of geological investigation is known as stratigraphy, in which an attempt is made to decipher the order in which the rocks were formed. Originally, the sequence of sedimentary rocks was determined by two methods. It was assumed that strata which were in sequence represented successive periods of time. Supposing that three sedimentary rock layers, *a*, *b*, and *c*, are found in the order shown in Fig. 12 (*b*), then *c* was laid down before *b* and *a* was laid down after *b*. In another area of the country stratum *c* might be found above the two other layers *d* and *e* (Fig. 12 (*c*)), which could therefore be regarded as of earlier origin than *c*. Let us suppose that in each of three distinct areas there are three strata (Fig. 12). In area (i) and area (ii) *a* is common to both, whilst in areas (ii) and (iii) *c* is a common rock. It is reasonable to assume that the complete

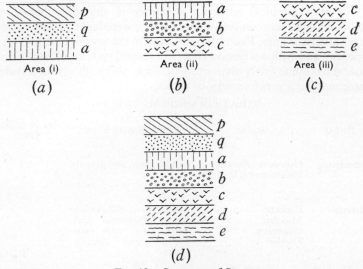

Area (i)

(*a*)

Area (ii)

(*b*)

Area (iii)

(*c*)

(*d*)

FIG. 12.—Sequence of Strata

sequence of rocks is as shown in Fig. 12 (*d*). In the early days of geological surveying, the recognition of this sequence depended upon the identification of rock *a* in two separate areas and so on. Whilst most sedimentary rocks have been laid down in water, it is possible for differences to occur in a given deposit over large distances, since the conditions under which deposition occurred could vary in different parts of the sea. Thus the outward appearance of the same stratum might be very different in two separate localities.

A surer method of checking the rock sequence involves the careful study of the fossil content of the rocks. To the layman two related fossils may look almost identical, but to the trained geologist certain subtle distinctions show that they are different. Such minute differences are the result of evolution, which produces marked changes in the structure of an organism over a long period of time.

Such changes could hardly be wrought in the period required for the deposition of one thin layer of rock, and so one particular fossil may be used to identify a given stratum, several feet in thickness.

In more recent times the geologist has also made use of igneous rocks as a means of dating the rocks. It has already been noted that there are within the crustal layers of the earth radioactive minerals which apart from creating heat, produce of their own accord new substances by the process of disintegration. The rate of disintegration of each mineral is constant, and the amount of new substances which has been produced is used to calculate the age of the rock.

The stratigraphical sequence is shown in the table below, together with the representative rocks which were laid down at various times and their ages. Each main division in the sequence is known as a period, and within a period there are several systems of rocks; each system has some fairly characteristic rock, although within a system there may be a great variety of rocks.

STRATIGRAPHICAL TABLE

Period	System	Representative Rocks	Age in Million Years
Quaternary	Holocene (Recent) Pleistocene (Glacial)	Sands, clays, and gravels	1
Tertiary	Pliocene Miocene Oligocene Eocene	Sands, clays, and gravels. Some limestones and some areas with coal (lignite) and igneous rocks	70
Secondary	Cretaceous Jurassic Triassic	Chalk Limestones and clays	200
	New Red Sandstone	Sandstones	
Primary	Permian		
	Carboniferous	Coal Measures Millstone Grit Carboniferous Limestone	
	Devonian (Old Red Sandstone)	Sandstones and limestones	
	Silurian Ordovician Cambrian	Mainly slates and shales with some igneous rocks	500
Archean	Pre-Cambrian	Mainly igneous and metamorphic rocks	About 2,000–2,500

This stratigraphical table is of geological importance, and each system or group of rocks within the major periods may include a great variety of sedimentary rocks and, since they have been formed at different times during this long interval, it is necessary to include igneous rocks as well. Without belittling the value of a fairly deep knowledge of the stratigraphical subdivisions, the primary knowledge which a geographer should gain is that of the relationship between the rocks and the landscape. Knowledge of the resistance, or lack of resistance, of certain rocks to the processes of weathering is thus desirable.

For instance, the association of many upland areas of Wales and Scotland with resistant rocks, chiefly slates of Primary age or metamorphic rocks of the granitic type, should be noted. Of course, the upstanding nature of such areas does not depend solely upon the

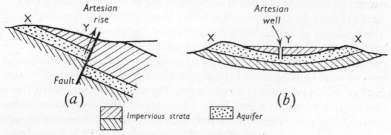

Fig. 13.—Artesian Water

In both diagrams the head of water at X in the saturated aquifer is higher than Y, thus permitting a rise of water to the surface along the fault in (a) or by the well which pierces the impervious layer above the aquifer in (b).

resistance of these rocks, since each area has a structural history which must be understood as well. In parts of south and eastern England chalk and limestone rocks, by virtue of their joints and pervious nature, are resistant to water erosion on their surfaces and form uplands in contrast to the neighbouring lowlands which lie on the less-resistant clays and sandstones of Secondary age.

Knowledge of the water-bearing properties of rocks is of obvious economic value. Sandstone and limestone, amongst other rocks, are very pervious, and are frequently completely saturated with water in their pore spaces or along joint planes, as compared with impervious clays and shales. Further discussion of this topic is given on p. 127. A stratum that bears much water, which is also free to move within the rock, is called an aquifer. Rock structures are of importance in relation to water supply. For instance, a fault may place an aquifer against an impervious stratum and cause underground water to rise to the surface (Fig. 13 (a)). This is an artesian rise of water, although artesian wells are more usually associated with downfolds (Fig. 13 (*b*)).

Coal and Petroleum

One other type of rock must be referred to, namely coal. Before coal can form there must have been a production of peat. Peat is derived from vegetation which has decayed under rather special conditions, so that the raw vegetable material has been humified. Further details of the production of humus will be given in a later chapter dealing with the formation of soils. For the moment it is sufficient to note that the plants and trees from which the peat is formed have grown in bogs and swamps, and many areas of the world show the active production of peat today. The Dismal Swamps of North Carolina (see Fig. 137) have modern peat layers as much as seven feet in thickness. Ideal conditions are probably found today in such areas as the Ganges delta, where there are forested swamps and a hot humid climate. With humification the plant tissues are converted into a residue which is rich in carbon as well as various hydrocarbons (compounds of carbon, hydrogen, and oxygen). Investigation of coal-bearing rocks show that there is a succession of seams each of which represents a peat layer. The seams are separated from one another by sand or clay beds. This suggests that the swamp areas in which the peat formed were deltaic, and suffered frequent changes in level over long periods of time. With depression the sea covered a peat bed, and on this were laid coarser sands or clays, until elevation gave a renewal of swamp conditions and the formation of a fresh peat layer. These overlaying beds of sand produced sufficient pressure to squeeze out of the peat both water and gases, and this increased the proportion of carbon in the material which could then be called a coal.

Various types of coal can be identified, and the varieties depend upon the plants from which the peat was formed, and the amount of chemical change which has occurred in the residue. In peat (and the majority of coals) the structures of the plants and trees are still visible. Lignite or brown coal is the least altered form of peat, and has a relatively low carbon content. Most lignites are of Tertiary age, but the valuable coals, which have a fairly high carbon content, are usually much older, being almost exclusively Carboniferous, or late Primary, in age. The result of greater chemical changes in these coals is to form bituminous types, which grade finally into anthracite, which is almost pure carbon. The removal of so much volatile hydrocarbons in anthracite is due to the effect of very great pressure having been exerted on the coal. These pressures are often related to past earth movements and a probable increase of temperature, so that anthracite may be a metamorphosed bituminous coal. Boghead coal and cannel coal seem to be derived from microscopic forms of vegetation (algæ) which have collected as very fine muds and oozes. They have less carbon than the other coals, but they may yield oil.

In passing it may be of value to draw attention to the precise significance of the strata known as the Coal Measures. These rocks may have a thickness of several thousand feet, and yet only some forty or fifty of these feet may be coal seams. This small proportion of coal is due to the fact that consecutive seams are separated by a varying thickness of deltaic sands and clays as described above. In some areas there are no seams of coal within the Coal Measures. It is important to interpret British geological maps correctly, since outcrops of Coal Measures are shown on these in black, and yet no coal seams may be present in some areas. There are coals and lignites in rocks of Secondary and Tertiary age.

Another mineral of organic origin found in rocks is petroleum, which consists of gaseous and liquid hydrocarbons whose source is probably both animal and vegetable. As a liquid, it can migrate through pervious rock. Petroleum is thus found in porous strata, or what are called reservoir rocks, since these are so capped and underlaid by impervious rocks that the liquid is actually sealed off as a pool. But the concentration is always too great to be entirely derived from organic remains in the immediate vicinity. An oil pool, we must realize, is therefore nothing less than a rock, like an aquifer, into whose pores oil has collected by migration from source rocks: some of these may be a considerable distance from the pool. In such concentrations, gas frequently occupies the pore spaces above, and groundwater lies below, the petroleum.

The rock structures associated with oil pools are various—in particular, we may note upfolds and strata that have arched into such forms as salt domes. The latter represent injections of salt into upper from lower strata under very great pressure, and they may be likened to laccoliths (see p. 23) except that no igneous activity is involved. Thus in Texas many oil pools are found on the flanks of these domes. Secondly, oil may be trapped in a reservoir rock by a fault that has brought this rock immediately against an impervious one. Here the structure is that of Fig. 13 (b) for artesian water, where the aquifer could be regarded as containing the oil pool. However, since most oilfields are found in folded rocks of Cretaceous and Tertiary age, their locations are closely related to the Alpine fold mountains shown in Figs. 45 and 46.

Isostasy

Evidence obtained from measurements of gravity in many areas of the world has led to the belief that the high parts of the continental masses are balanced by a sial layer which is somewhat thicker than elsewhere. The sial is less dense than the sima, but a balance is struck which may be illustrated as in Fig. 14. It has been suggested that the continents are rafts of sial 'floating' in a layer of sima, the substratum. If this view is permissible, then Fig. 14 represents a solid continent floating in a 'liquid'. Hence, at a depth indicated by the line XY, there should be equal pressures on equal areas of a plane lying at that depth. In other words, if four columns of equal cross-section (1, 2, 3, and 4) be considered, then the weight of material

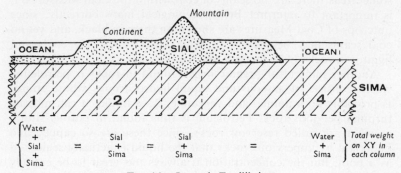

FIG. 14.—Isostatic Equilibrium

in each column should be the same. Thus the thin, light sial layer of column 1 must be balanced by a thick, dense sima layer, whilst a thick sial layer as shown in column 3 must be balanced by a relatively thin layer of sima and so forth. If this balance exists, then the various columns are said to be in isostatic equilibrium with one another. The word isostasy means 'equal poise', and therefore upstanding surface areas and low-lying areas are adjusted with regard to one another.

It is very unlikely that isostatic equilibrium is maintained for very long. Let us see how it may be disturbed, and then how a readjustment is possibly made. In Fig. 15 part of the cross-section shown in Fig. 14 is considered. Let us assume, for the sake of our argument, that erosion has removed the top of a mountain and spread the rock waste to a uniform depth over the continental surface at the top of column 2. This has disturbed the isostatic equilibrium, for clearly the base of this column is now supporting a greater weight than the base of column 3 which has now a lighter load. There is too great a thickness of sial in column 2 and too little in column 3 for equilibrium to exist. Recalling that the sima may at times become fluid, it is regarded as possible that some of the sima flows from column 2 to column 3. The removal of sima from column 2 would correct the overloading, whilst the addition of sima to column 3 would make up the deficiency of load therein. This would result in an elevation of the sial below the mountain region and a depression of the sial of the continental surface nearby. Since it is absurd to think of the sial of the continent moving up and down in separate columns, the adjust-

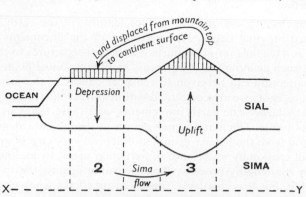

FIG. 15.—Effect of Disturbance at Surface and readjustment of Isostatic Equilibrium

ment is probably effected by a warping or tilting of the sial block, so that one part of its surface rises and another descends. Such warping could be responsible for marine transgressions which have been referred to on p. 10.

Continental drift is not explained by isostasy, but the 'flotation' of the sial rafts is dependent on this equilibrium. Large-scale movements of the deeper sima, at the time when isostatic adjustments are taking place, may be responsible for some lateral shifting of the continents. Later in this book it will be necessary to refer to isostatic uplift and depression in order to explain certain phenomena such as the formation of mountains from geosynclines (p. 39) and various glacial problems in Chapter VI.

CHAPTER II

MAJOR STRUCTURES OF THE EARTH—
(1) VOLCANIC LAND FORMS

Introduction

In the ensuing sections the surface features of the earth will be examined in some detail. A description of the lineaments of the earth's surface falls into two parts. The major structures of the continents, such as mountains, plateaus, and plains, must be discussed before attention can be given to the finer details of the picture. In the first category we are concerned with features that are determined mainly by large-scale earth movements, and these range over vast periods of time. The many minor details of the second category are developed from the former group, and depend very largely upon the effects of erosion by water and, in particular areas, by ice and wind. Erosion is a lengthy process when measured against the span of a man's life, but it is relatively speedy in its effects when compared with the time involved in the production of a mountain range. It is frequently necessary to consider the effects of large-scale earth movements as well as the later and more recent effects of erosion when describing the details of a particular land form. In the discussion that follows, the major pattern of the earth's surface will be traced by reference to vulcanicity, mountain building, and plateau construction, with associated rift valleys. The decoration of this pattern by the minor details will be developed later from a study of the work of rivers, ice, and wind.

Vulcanicity

Below the crustal layers of sial and sima there exists material which periodically becomes liquid, or at least plastic, and is forced upwards to the surface of the earth. Despite the very high temperatures which prevail at these depths, this material is more frequently solid, and this state is maintained largely by the great pressure exerted by the superior layers. Locally at depth an increase of temperature, combined with a decrease of pressure, will cause this material, together with the basal layers of the sima, to become molten rock or magma. Magma is largely mineral material, but it includes much gas. On becoming plastic or molten, it rises along conveniently placed fissures to the surface of the earth. Sometimes it invades only the lower portions of the crustal layers by a process of engulfment of masses of solid rock, and its upward movement is arrested before the surface is reached. Access to the continental surfaces or to the ocean floors produces volcanic activity. The pressure upon the magma decreases as it ascends from below, and gases and other volatile substances are given off, so that the magma reaches the surface with relatively little gas enclosed within it, and consists almost entirely of mineral material. This mineral residue eventually chills and forms solid rock, composed very largely of various mineral crystals.

When the magma chills inside the crustal layers, it produces several types of rock masses (Fig. 16). Great masses known as bathyliths, with domed upper surfaces, may extend upwards from

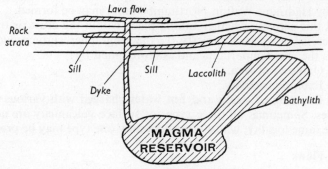

FIG. 16.—Forms of Igneous Rocks

a deep-seated magma reservoir. At higher levels and nearer the surface the molten rock may have been squeezed in between two layers of surface material, and has produced a laccolith on cooling. The supply of magma to a laccolith has possibly been effected by means of a thin horizontal layer of molten rock which was an offshoot of a vertical stream of molten rock. These horizontal and vertical layers of chilled rock are known as sills and dykes respectively. The injection of rock to produce a laccolith has distorted the surface layers above, and gives a dome-shaped mountain or hill mass. Such are found in the Henry Mountains of Utah, in the upper part of the Colorado Basin. Removal of the surface layers by erosion may expose the surface of bathyliths and laccoliths. This has occurred in South-west England, where Dartmoor, Bodmin Moor and associated uplands are representative of these exposed surfaces. Phacoliths are similar to laccoliths, but the magma has been injected into folded structures. If the magma collects on the crest of an upfold, then a saddle reef is formed, which is frequently the site of mineral deposits of value.

When the magma flows out on the surface of the earth, it is known as lava, which spreads in extensive sheets and, in contact with the atmosphere, chills fairly rapidly to form a lava flow. Extrusion of lava is not infrequently accompanied by explosions, which help to build up volcanic cones. Broadly speaking, the rock which solidifies in depth, in the crust, is coarser in grain than the rock which forms on the surface. The rock in bathyliths and many laccoliths is granite or akin to granite, with large crystals. Lava flows solidify to produce a fine-textured rock with small crystals known as basalt. The rocks in sills and dykes are usually not unlike basalt (see p. 13). When weathered and exposed, sills and dykes may give rise to local relief features. Thus waterfalls or a narrow ridge, such as that followed in part by Hadrian's Wall in Northumberland, may be formed.

All this activity, whether taking place within the crust or on its surface, is known as vulcanicity. To the geographer, surface vulcanicity is the more important, and is represented by:

(a) Outflows of lava.

(b) Explosive activity.

(c) Emission of steam and hot water charged with various substances. Sometimes all three types of surface vulcanicity are active in the same locality, whilst elsewhere only one type may be present.

Lava Flows

Lava in these flows is not usually accompanied by large quantities of gas, so that it represents a relatively gentle form of volcanic activity. Today lava may be seen pouring out periodically in Iceland. It issues spasmodically in great flows from a long line of fissures. Detailed geological study of areas, such as the North-west Deccan of India and the plateau of the Columbia-Snake Rivers in the states of Washington and Oregon, show that in the past there were many episodes of lava outpouring, each possibly separated by very considerable intervals of time. The explanation of this intermittent activity is probably related to the conditions at depth which control magma movements. It is suggested that the rock material becomes fluid in only small quantities at a time. The extent to which lava flows from a fissure, before chilling is sufficient to cause it to become solid, is determined not so much by its initial temperature as by its mineral composition. If the lava is rich in minerals containing silica, it tends to be viscous, and so its progress is limited. On the other hand, when the silica content of the lava is relatively low and the ferro-magnesian minerals are abundant, it is very mobile, and the lava will extend over great distances before solidification occurs. In all cases the cooling is quite rapid, and only small crystals of minerals are formed. Excessively rapid chilling may result in the production of volcanic glass in which no crystalline structure may be discerned.

Lava Cones

Not always does the lava find its outlet by a fissure of great length. Eruption of lava is often confined to a small vent, and it is then that a cone develops. Usually this restricted vent is responsible for encouraging explosive activity, since it may become sealed by chilled rock, which must be blown away before further activity is possible. There are various types of cone, and each is determined by differences in the lava and by the presence or absence of vigorous explosive activity. Some cones are built almost exclusively by lava, which flows out from a crustal vent. A very mobile lava will develop a low cone, which is little more than a series of lava flows, and which is higher in the middle and radiates in all directions from the vent (Fig. 17). Such are known as shield volcanoes.

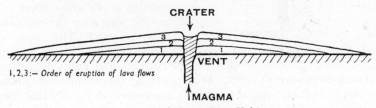

CRATER

3   3
2   2
1   1

VENT

1,2,3:— Order of eruption of lava flows

MAGMA

FIG. 17.—Basic Lava Cone Volcano

Successive outpourings of lava will tend to build up the cone to a certain extent, but such cones invariably show a base which is broad in relation to the height. One of the Hawaiian volcanoes, Mauna Loa, which is derived from very mobile lavas, measures over 70 miles across at the base, and yet it is only about $2\frac{1}{2}$ miles above sea-level at the top of the cone. The base is, in fact, on the sea floor.

On the other hand, a viscous lava will produce a dome-shaped cone, since the congealing of the lava in this case is quite rapid and the most elevated part has a height which is great in proportion to the breadth of the base. Further additions of lava in such cones develop from inside, and the earlier-formed layers are heaved up (Fig. 18). Extinct cones of this type are to be found in the region of Auvergne in Central France, where they help to build the characteristic landscape of the area (puys d'Auvergne).

Explosive Activity

As the magma rises towards the surface, there is expansion of its included gases, and this may give explosive activity. This causes a fragmentation of the lava much in the way that air bubbles shatter the surface of boiling water. It has been noted that a sealed vent must be blown free by renewed activity, so that explosions may give a shower of fragments consisting of both new lava and previously formed rock. The fragments which are ejected by these explosions may be fairly massive, volcanic bombs as large as big boulders, but, in general, they are smaller, and may decrease in size to fine particles comparable to dust. The larger fragments usually fall around the top of the vent or the crater, and assist in building the cone upwards, whilst smaller fragments are responsible for developing the

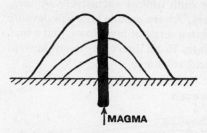

↑MAGMA

Fig. 18.—Acid Lava Cone Volcano

cone outwards, since the latter are thrown farther and fall on the sides of the cone. The finest of this fragmentary material is known as volcanic dust, and often remains suspended in the atmosphere for a long time after the eruption has occurred. This dust is usually carried away by air currents and gradually settles on areas of the earth's surface or on the ocean floors which are remote from the volcano. Some fragments are cooled so rapidly that they form volcanic glass, without any crystalline structure. When large quantities of this dust are disseminated through the atmosphere, particularly vivid sunsets are common. After eruptions of great magnitude cool weather may be a feature of the ensuing months, due, it is thought, to the interception of the sun's rays by the fragments in suspension in the air and the consequent reduction of insolation at the earth's surface.

The alternation of lava flow and explosive activity creates a composite volcanic cone which is typical of most volcanoes. The structure of a composite cone is shown in Fig. 19. It will be observed that the cone is relatively high in comparison to the lava cone. The angle of the slope of the sides of the cone is usually about 30°, being determined by the angle of rest of the fragmentary material. Each successive layer of lava extends the base of the cone in a lateral direction. Lava has, at certain stages in the history of the volcano, issued sideways, with the formation of dykes which have strengthened the cone. Sometimes active conelets are established on the flanks of the main volcano; these are known as parasitic cones. Vesuvius is an excellent example of a cone built of lava flows and fragmentary material. In some instances eruption has been exclusively of the explosive type, so that the cone is composed entirely of fragmentary material. These

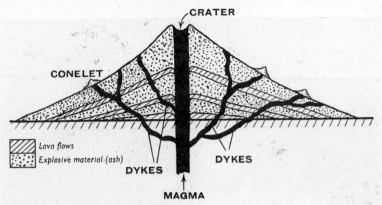

FIG. 19.—Composite Cone Volcano

cones are lofty, and resemble in appearance the composite cones. Monte Nuovo, near Naples, is an example of such a cone, also El Misti, in Peru (Fig. 20).

Catastrophic explosions which come after a period of quiescence may result in the structure of a volcano being radically modified. The upper part of the cone is weakened to such an extent by the explosions that it subsides, the normal crater at the top of the vent becoming much wider and so a caldera is formed. Fracturing of the sides of the caldera is quite common (Fig. 21). Kilauea, in the Hawaiian Islands, is a caldera in one of the largest lava cones in that region. Krakatoa, in the Sunda Straits, was converted from an island volcano into a great caldera which is now largely occupied by the sea as a result of a very violent eruption in 1883. The cone of Vesuvius was wrecked by the explosive eruption of A.D. 79, when Herculaneum and Pompeii were overwhelmed by the lava flows which succeeded the eruption. Frequently a new cone arises within a caldera. The modern cone of Vesuvius is placed inside the older caldera, and in this century Anak Krakatoa (the Child of Krakatoa) has appeared in the sea area which occupies the site of the earlier island.

Volcanoes may be distinguished by the rhythm and intensity of their explosive activity. The fairly regular and moderate explosions which are characteristic of Stromboli, in the Lipari Islands of the Tyrrhenian Sea, are replaced in the case of Vesuvius by irregular and violentactivity,accompanied by a towering cloud of fragments. Sometimes the upward release of gases is impossible on account of a plug of chilled lava which seals the vent, and then the gases and molten

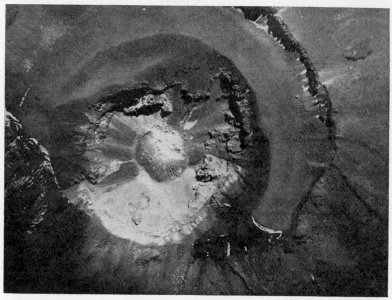

James Sawders—Combine

FIG. 20.—El Misti, Peru

A volcanic cone of fragmentary material, showing the crater from the air.

lava issue sideways as a blast of hot fragments. These blasts are made very mobile by the abundance of gas, which helps to reduce friction between the moving fragments. Montagne Pelée, on the island of

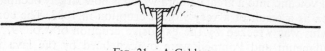

FIG. 21.—A Caldera

Martinique, erupted in 1902 by the uplift of a spine of chilled lava which permitted the side issue of such *nuées ardentes*. This type of eruption is probably the most dangerous of all, since nothing which is in the path of the erupted material is likely to escape engulfment. The town of St. Pierre, which lay below Montagne Pelée, was therefore literally smothered in hot dust.

Land Forms produced by Volcanoes

Volcanoes may form quite notable elements in the structure of the continents, and many significant mountain peaks in the Andes, the western mountains of North America, the islands off eastern Asia, and in many other localities are volcanic cones (Fig. 22).

V. C. Browne

FIG. 22.—Mount Egmont, North Island, New Zealand
A typical volcanic cone, showing both radial and dendritic drainage.

Some are active today, but the formation of chilled lava in the vent of a volcano which may extend down as a solid plug below the base of the cone produces either an extinct or a dormant volcano. Dormancy may be ended, as in the case described above of Montagne Pelée, when the chilled plug is forced out by renewed activity. In extinct volcanoes there has been no activity over long periods of geological time. The result is that the cones of such volcanoes have been reduced by erosion so that their conical shape is either much modified (see p. 80, radial drainage-planèzes) or largely destroyed. The almost complete reduction of an extinct cone is seen in the volcanic crags, such as those of Arthur's Seat and the rock on which the castle is situated, at Edinburgh. Here the outer parts of the cone have been removed, but the solid plug which sealed the volcano has been more successful in resisting erosion, so that it remains as a conspicuous feature of the landscape (Fig. 23). In the

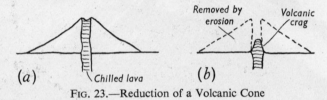

(a)   Chilled lava   (b)

Fig. 23.—Reduction of a Volcanic Cone

craters of some extinct volcanoes there are lakes today. Several of these are found near Naples, in the Phlegrean Fields, and others are located in the Eifel district of the Rhineland of Germany. The most famous example is that of Crater Lake in Oregon. A crater lake is shown in Fig. 24.

The distribution of the outstanding active volcanoes in the world today is given on the maps (Figs. 45 and 46).

Geysers and Hot Springs

A relatively mild form of vulcanicity is the emission of steam and hot water charged with various substances. Such activity is fairly restricted, and occurs mainly in areas where vulcanicity has declined recently, from a geological point of view, although it is sometimes associated with volcanic areas which are still active. The most common gas which is given off by rising magma is steam, but certainly some atmospheric water which has seeped downwards must be considered as contributing to these emanations. Fumaroles and solfataras are points of issue, on the sides of volcanic cones and on lava flows, for much water and steam. Quite frequently silica and calcium carbonate are deposited around these springs to form fantastic, coloured layers of siliceous sinter and travertine respectively. Such deposits give the terraces of the Yellowstone National Park in Wyoming and of certain areas of North Island, New Zealand. At no great depth pressure is so increased that water will not boil until it has been heated by steam rising from hot magma far below, to a temperature which is much above the normal boiling-point. A sudden decrease of pressure will have the effect of changing, instantaneously, such

The Aircraft Operating Co. of Africa (Pty.) Ltd.

FIG. 24.—Crater Lake, near Lake Kivu, Belgian Congo

Part of the fault scarps of the western East African rift valley is seen in the background.

superheated water into steam. Geysers are the evidence of this particular activity, and they are intermittently active because they depend upon a sudden release of steam in this way. Much water is then hurled from relatively shallow reservoirs which lie below the surface of the earth, with such force that it may rise in a great spout to a height of 100 or 200 feet in the air. Each ejection of water is preceded by a period of quiescence, during which the temperature is gradually mounting to create a new paroxysm of boiling water under great pressure. Geysers are not widespread throughout the world; they are found chiefly in Iceland, New Zealand, and the Yellowstone National Park in the United States of America.

CHAPTER III

MAJOR STRUCTURES OF THE EARTH—
(2) MOUNTAINS

Mountains

Every continent has a variety of structural features which give it a distinctive pattern. A relief map of any continent will show very clearly the contrast between highlands and lowlands, so that the recognition of many mountains is relatively simple. From the point of view of the geologist and the geomorphologist, a mountain region is not determined solely by its superior relief, but also by its structural characteristics. It may happen that regions which display these structures are relatively lowland areas today. The predominant features which may therefore be regarded as typical of mountains are the product of certain geological processes.

Folding

In Chapter I the effect of compressional forces acting tangentially along the surface of the earth which produce simple folding was observed. Although such simplicity of structure may be seen in

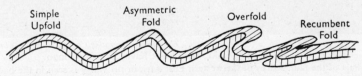

FIG. 25.—Types of Fold

certain mountains, it is usual to find that various elaborations are common, so that overfolds and recumbent folds are developed (Figs. 25 and 31).

Faulting

Faulting has already been mentioned in an earlier section of this book. The forces that lead to this vertical displacement of rock masses are essentially in directions at right angles to those which produce folding. They are thus radial to the surface of the earth. Reference to Fig. 26 will show the fault pattern of the Appalachian

32

Mountains. The faults are seen to cut across the fold pattern, which suggests that faulting may follow earlier folding in the development of the structure.

Thrusting

Faulting may be developed in a different plane from the vertical, and the forces at work seem to have acted in such a way that the relative displacements of rock masses are more nearly horizontal.

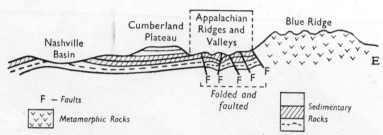

FIG. 26.—Section across the Appalachian Mountains

These forces act along a thrust plane, and they result in the movement of rock masses over very considerable distances. Within many mountain zones several thrust planes have developed together with intermediate minor thrust planes which are usually inclined to the major ones. The complex structure which has developed in this case is called imbricate structure, and it is associated particularly with the North-west Highlands of Scotland (Figs. 27 and 28). When

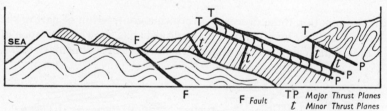

FIG. 27.—Imbricate Structure in North-west Highlands of Scotland

thrusting has developed along the limbs of recumbent folds, the displaced strata form nappes. Such overthrust masses of rock are a particular feature of the Central Alps of Europe (Figs. 29 and 97).

Stages in Mountain Formation

There is clearly a relationship between these various processes, since they all tend to produce a reduction in the surface area of the

P.G.—3

rock masses. This shortening of the surface is regarded as an essential feature in the building of mountains, and the sections of various mountain regions shown in this chapter demonstrate that singly or

Geological Survey photograph, reproduced with the permission of the Controller of H.M. Stationery Office

FIG. 28.—Moine Thrust Plane, Knockan Cliff, Sutherlandshire

Pre-Cambrian Moine schists (above) have been thrust over light-coloured Cambrian limestone (below). The thrust plane lies just above the hammer head.

in combination these processes produce a contraction of part of the earth's surface. Investigation by geologists has shown that much of a mountain region is built of great thicknesses of sedimentary rocks,

FIG. 29.—Formation of a Nappe

some being very considerably metamorphosed. Where a particular group of sedimentary rocks can be traced throughout a mountain zone and beyond it into the adjacent lowlands, it frequently happens that the thickness of these layers varies considerably. For instance, certain strata are 25,000–40,000 feet thick within the Appalachian

Mountains, and yet the same group of rocks in the Mississippi Basin, to the west, is only 4,000 feet thick. Such a thickness as 40,000 feet represents a long period of accumulation, and it would appear that the mountain zone has been derived from the section of the deposits which was thickest.

Intense folding and thrusting, together with faulting in depth, have given access to molten rock from below. Thus the mountain structures are ramified by masses of igneous rock which have resulted from widespread vulcanicity. In some mountain regions active volcanoes are found, and a good many mountain areas are characterized by the remnants of basaltic lava flows and volcanic cones, whilst within the cores of the intensely folded structures are granitic masses which represent the exposed upper surfaces of great bathyliths (see Fig. 32). Metamorphism of the sedimentaries, and even the igneous rocks, has been inevitable owing to the enormous strains due to the thrusting, combined with the effects produced by the contact of molten rock. These effects have been so widespread that they are known collectively as regional metamorphism. Since metamorphism involves a rearrangement of the minerals in a rock and, often, a molecular rearrangement, there have developed fresh minerals. It is, perhaps, not surprising that some mountain areas are important sources of various metallic ores.

Pattern of Mountain Regions

Mountain regions of the world today may be grouped according to their geological age. It is apparent that some mountains are 'young', whilst others are 'old'[1]. The geologist is able to determine the age of the various rock layers which have been involved in the building of a mountain area. Clearly the presence or the absence of rocks of a given age is a clue to the time of mountain building. Admittedly, the dating of a period of mountain construction is not quite so simple as this but, in principle, this is the main line of approach to the problem. Other lines of evidence are employed, but a discussion of these is beyond the scope of this book. Obviously 'older' mountains have had more time to be affected by the process of denudation, so that although the typical mountain structures are present there are substantial differences in the appearance of these mountains as compared with examples of 'younger' mountains. Although the 'younger' mountains have been affected by denudation, they usually have a greater height than the 'older' ones which have frequently lost the irregularity of surface that was determined by the

[1] The terms 'young' and 'old', are used in the relative sense since even 'young' mountains may be several million years in age.

original folding. Prolonged denudation has reduced them to pene-plains (almost level surfaces at a low elevation), or they stand out as plateau regions, which are peneplains that have been elevated by uplift. It is sensible, then, to examine the pattern of 'young' moun-tains, since they are more likely to show the ground plan of what may be regarded as a typical mountain zone. This pattern is related to 'older' mountain areas that are represented by plateaus bordering the 'younger' mountain region. The simplified pattern may be shown by a diagram (Fig. 30 (*a*)). Highly folded (Fig. 31) and thrusted zones of sedimentary rocks lie between areas which are remnants of 'older' mountain blocks. The mountain zone is invariably longer

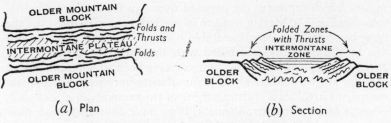

(*a*) Plan   (*b*) Section

FIG. 30.—Pattern of a Fold Mountain Region

than it is wide, and included within the folded region is an area of relatively undisturbed rocks which are recognized as intermontane plateaus (Figs. 30 (*b*) and 234).

The ideal plan, as shown in Fig. 30, is readily matched in all its details in any of the present-day mountains. Perhaps the nearest approach to the ideal ground plan is afforded by the extensive systems of mountains which cross the great land mass of Eurasia, including the Alps and their associated ranges, as well as the Hima-layan region. Fig. 45 will show the two border regions of plateaus which are 'older' mountains. These are represented by most of northern Africa, Arabia, and the Indian Deccan on the south side, whilst on the northern side is the series of plateaus and plains in Central and Northern Europe and in Central Asia. The folded and thrusted zones are shown by a series of sinuous lines enclosing such intermontane plateaus as Tibet, Iran, Anatolia. Farther west in the Alpine region these plateaus are not represented, but in their place are the basins of the Mediterranean Sea which have been produced by extensive faulting followed by foundering of intermontane areas. The great western mountains of the Americas exhibit a similar, if apparently less complete, plan. In North America the eastern folded and thrusted zone of the Rocky Mountains and the western zone of the Cascade Mountains, Sierra Nevada, and the several coast ranges are separated by the intermontane plateaus of Colorado and the Great Basin of Nevada, with the adjacent lava plateaus of the Columbia-Snake River basin. In Central America the outer folded zones are literally one, although an intermontane region is

The London Electrotype Agency Ltd.

FIG. 31.—Allegheny Ridges and Valleys, United States of America

Denuded fold structures in the Appalachian Mountains. Note the trellised drainage which is exhibited.

found in Mexico. South America has in the Andean region only one extensive plateau area which is restricted to parts of Peru and Bolivia; elsewhere the two folded zones have coalesced to form one, although smaller intermontane areas are recognized in the northern parts. These American mountains have bordering blocks of 'older' mountains in the east of Canada and the United States and in South

America, but elsewhere in the west and to the east of Central America these blocks have foundered, and lie in the Pacific and Atlantic ocean basins, in all probability.

The ground plan of certain mountains shows a pattern of mountain arcs. These are very characteristic of the folded mountains which thread the islands off the eastern coast of Asia. The trend of a mountain range in one island may be clearly traced into an adjacent island. The arcs are convex towards the Pacific Ocean and similar curves are noticed in the Himalaya Mountains and on the south side of the Iranian Plateau (see Fig. 45). It is possible to trace similar arcs in other mountains such as the Carpathian Mountains and the Andes Mountains, but they are isolated and do not form a series of festoons of arcs as in Asia. It has been suggested that these Asiatic arcs are the surface outcrop of great thrust planes which extend inwards from the surface of the earth and towards the interior of the continent of Asia. Each arc may be related to a 'pole', and it has been shown that the poles of the Asiatic arcs align themselves along parts of two great circles.

The Process of Mountain Building

In the preceding pages a description of the structure, rocks, and the ground plan of mountain regions has been attempted. Many theories have been advanced to give an adequate explanation of the piling of rock masses into a mountain system. Early attempts were concerned with the contraction of the surface layers of the earth as cooling down proceeded through long periods of geological history. Jeffreys[1] computed the loss of surface area which would occur as the result of a calculated contraction of the earth. Unfortunately, a re-examination of his calculations showed a great discrepancy between the shrinkage that Jeffreys allowed and the reduction of surface area required by the geologist to produce all the older and younger mountains which were recognized. Joly[2] developed a theory which turned upon the accumulation of radioactive heat within the rocks of the sima. He argued that periodically sufficient internal heat accumulates in the sima to make it plastic, and thus allows the sial rafts to subside to a certain extent. This subsidence leads to an extension of the oceans, whose waters would in time cause a dissipation of this heat, and then the sima gradually returns to a more rigid state. The contracting sima would compress the sial rafts from the sides and also thrust them upwards. The combined effect would, in

[1] Harold Jeffreys, Plumian Professor of Astronomy and Experimental Philosophy at Cambridge.
[2] John Joly, Fellow of Trinity College, Dublin.

Joly's opinion, develop the structures of folded mountains. Neither of these two theories, which have been merely outlined, is acceptable today, although each may contain an element of the truth. No satisfactory explanation has yet been given of the forces which are required to build mountains, but although modern theories are very divergent in many of their details, there is general agreement that a geosyncline must be formed prior to mountain development.

The geosyncline is regarded as a long and narrow depression which developed in the sialic crust, and was occupied by the sea. In it sedimentary rocks, derived from continental areas, were laid. It will be remembered that observation has shown that parts of the Appalachian Mountains are built of rocks which are some ten times thicker than they are outside the mountain region. Accumulation of such a thickness of rock would only be possible if the bottom of the geosyncline was steadily sinking throughout the very long period of deposition. Forty thousand feet of sediments in the Appalachian zone were accumulated over a period covering 100 to 200 million years. After a certain stage of deposition had been reached, the geosyncline would be overweighted, and from the simple considerations of isostasy (p. 20) adjustments in the load of the column of rocks below the geosyncline would occur. It is thought that the sial layer bends downwards beneath the geosyncline and sima is displaced. To compensate for this disturbance of the sima it is conceivable that the continents would be slightly uplifted. Continuation of this overloading of the geosyncline leads to the development of forces which act from the sides of the trough of sediments towards the centre (Fig. 32). In time the sediments are literally crushed between the continental blocks, which resemble the jaws of a vice, and intense folding of the rocks occurs. At the same time an uplift of the sediments, with thrusting (Fig. 33), results in the development of the other features which are common to mountains, such as the overthrusting of nappes, vulcanicity in its various forms, and the formation of intermontane zones which are parts of the geosynclinal deposits that have been relatively undisturbed by these forces.

An elaboration of this theory of mountain building from a geosyncline suggests that the sialic base of the geosyncline becomes so strained with down-warping that it fractures, and a great mass of sial sinks into the underlying sima. The gap which is thus produced permits the movement of the continental blocks on either side of the geosyncline with consequent folding. At the same time this gap gives molten magma easy access to the folded mass. Fig. 32 shows the stages of development of the Alps, based on such theories as have been outlined. It should be emphasized that in the stages which follow the crushing of the geosynclinal sediments, the elevation of the folded structures inevitably upsets isostatic balance and further adjustments are necessary. Since the sediments are relatively light, there is probably a tendency for sima to move back beneath the geosyncline and buoy them up. This uplift is no doubt responsible for the forces which develop thrusts, and the outward facing thrust

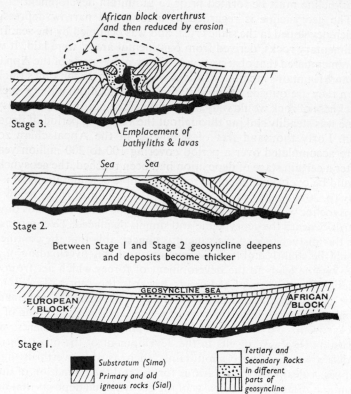

African block overthrust and then reduced by erosion

Stage 3.

Emplacement of bathyliths & lavas

Sea   Sea

Stage 2.

Between Stage 1 and Stage 2 geosyncline deepens and deposits become thicker

GEOSYNCLINE SEA

EUROPEAN BLOCK   AFRICAN BLOCK

Stage 1.

■ Substratum (Sima)

⧄ Primary and old igneous rocks (Sial)

Tertiary and Secondary Rocks in different parts of geosyncline

FIG. 32.—Alpine Mountain building in Europe

planes, shown in the cross-section of the Himalaya Mountains (Fig. 34), are probably due to this process. Geosynclines take a very long time to develop, and since there is a very considerable geological interval between mountains of different ages, it would seem that there are periods when geosynclines are absent from the surface of the earth. Today there is only one region which has some of the characteristics of a geosyncline, and it lies in the area of the East Indies and close to some of the mountain arcs which have been

Fig. 33.—Part of the Swiss Alps to the east of Zurich Lake

Upthrust sedimentary rocks are seen on the northern edge of the fold mountain zone. The view is taken looking eastwards.

mentioned. Off-shore to the south of Java and Sumatra the force of gravity is much less than normal. This observation has been interpreted as an indication of a down-warp in the sial which must be occupied by lighter rocks, presumably sediments. If this is true, then the mountains of the Asiatic arcs are very young indeed, and may, in fact, represent the earliest upfolded structures to emerge from a geosyncline.

Holmes[1] has developed a theory which seeks to explain the down-warping of the sialic crust more satisfactorily than perhaps the

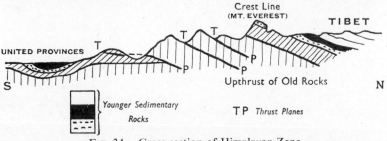

Fig. 34.—Cross-section of Himalayan Zone

Arthur Holmes, Regius Professor of Geology, University of Edinburgh.

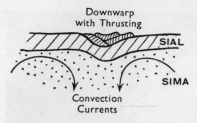

FIG. 35.—Downwarping of a Geoscyncline

earlier theories do. He envisages the development of radioactive heat in the sima layer which sets up currents in the plastic sima that are analogous to convection currents in a liquid. These may become so forceful that they drag the base of the sial downwards and create a geosynclinal depression (Fig. 35), in which eventually the sediments are crushed. Like Joly, he insists upon the dissipation of this heat after considerable time has elapsed. Then the convection effects wane and the sima chills and produces uplift in the mountain zone.

CHAPTER IV

MAJOR STRUCTURES OF THE EARTH—
(3) PLATEAUS AND RIFT VALLEYS—
CONCLUSION

Fault Structures

In the preceding sections it has been necessary to refer to the effects of both folding and faulting, together with thrusting, which have worked severally to produce mountain structures. It will be recalled that the major movements that are involved in such developments are due to forces which act more or less horizontally, and these may be called collectively, orogenetic, or in other words, they create mountains. In some areas the present structures may show that the predominant forces which have operated on the surface layers of the earth have been at right angles to the direction of these

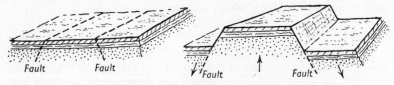

Fig. 36.—Development of a Horst

orogenetic forces, and the main effect is not to reduce the surface area by crumpling, but to produce a relative uplift or depression of adjacent areas in conjunction with large-scale faulting. The units of surface relief which are thus developed are known as horsts and rift valleys. Figs. 36 and 37 will show these two units in perspective. The classic explanation of the development of these features, which are almost inversions of one another, suggests that they are due to uplift or depression, or to both these processes acting simultaneously (the arrows in Figs. 36 and 37 show how upward and downward movements can account for these features). Whilst a horst may stand out above relative lowland on either side, a rift valley is largely defined by the two horsts which are its margins. In both cases the boundaries of these features of relief are steep edges which are known as fault scarps[1] (Fig. 43). Some of the most perplexing problems have been

[1] See footnote on p. 56.

involved in the explanation of the uplift of great blocks of the surface layers of the earth and the subsidence of sections of the crust to account for the formation of a rift valley. It will be noted, from Fig. 37, that the depressed block which has given the floor of the rift valley is comparable to the keystone of an arch, but it has sunk, despite the wedging effect of the horsts on either side. It has been suggested that the collapse of the crust wedged the horsts apart and often vulcanicity has occurred at the site of the faults at the foot of

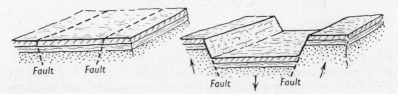

FIG. 37.—Development of a Rift Valley

the fault scarps, although this is by no means an inevitable consequence. At one time this collapse was considered to be a result of the contraction of the surface of a cooling earth. It has also been suggested, however, that tension in the earth's crust pulled the horsts away from the rift block which was then able to sink.

Plateaus

Since horsts, which may be extensive enough to produce great plateaus as major features of the earth's surface, can be regarded as products of uplift, the forces which are responsible for them are known as epeirogenetic. The expression epeirogenetic refers to the formation of the continental masses. It will not have escaped notice that one stage of mountain building involves extensive uplift, with the production of intermontane plateaus; however, uplift may occur without the accessory disturbances of mountain building. Many of the older mountains may be regarded as upland features which have been raised by epeirogenetic forces which operated much later than the orogenetic forces. Many of the Central European massifs, such as Bohemia, the Central Plateau of France, and also the North and Central Pennine Uplands, have been produced by such uplifts, although they may contain within their limits folded structures dating from earlier episodes of mountain building. The Highlands of Scotland are similar, only the plateau has been dissected by faulting and ice action (Figs. 38).

The 'older' mountains have been called block mountains by many writers, but it is more suitable to apply this term to more restricted

Geological Survey photograph, reproduced with the permission of the Controller of H.M. Stationery Office

Fig. 38.—Part of the Forest of Mamore, Inverness-shire

A dissected plateau of uplifted 'old' mountains showing the general plateau level in the background and glacial features with corries, arêtes, and U-shaped valley with a meandering stream in the foreground.

areas of uplifted territory which often show a distinct tilt, due to the irregularity of the forces of uplift, and which are bounded by distinct fault scarps. Tilted blocks of this type are found in the North and Central Pennine regions. The north part of this zone is bounded on its northern and western sides by faults, the latter giving the bold fault scarp of Cross Fell Edge. In section this region is seen to be tilted eastwards towards Co. Durham (Fig. 39). This tilted plateau of

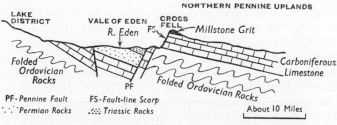

Fig. 39.—Tilted Block Plateau of North Pennine Uplands

northern England is a small-scale example of such extensive areas as the Deccan of India, the Arabian Plateau, the Brazilian Plateau, and many others. Such plateaus are widely distributed. When faulting is complex and dislocations of the surface are unequal, basin and range structure is developed as in the intermontane zone of the western mountain system of the United States of America. Surface patterns of faults are very intricate, and since both upward and

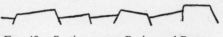

FIG. 40.—Section across Basins and Ranges
Region (U.S.A.)

downward displacements have occurred (although the main effect has been one of uplift), a series of plateaus or longitudinal flat-topped ranges are found in juxtaposition to depressed areas or basins. Most of the ranges are individually like the tilted blocks of the North Pennine region (Fig. 40). It is important to note that this section is in one direction, across the basins and ranges, and therefore emphasizes the tilting in that direction only. The floor of each basin and the surface of each plateau are tilted in a longitudinal direction as well. Fault developments, at right angles to one another, may subdivide a basin area into a series of minor basins each at a different level. Quite frequently tilts are so varied that certain basins are 'landlocked' by fault scarps on all sides, and have, in consequence, no

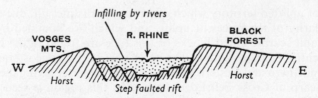

FIG. 41.—Section of the Rhine Rift Valley

outward drainage. These areas are often elevated and shut in by higher mountain terrain, so that they are largely deprived of rainfall. Such flows of surface water that may occur are inadequate to fill the basins and so produce lakes which could spill over the edges of the basin. Thus these basins are characterized by internal drainage with salt lakes and playas (Fig. 119). The plateaus of inner Asia are very similar, with enclosed basins, such as that of Tarim. It will be noted from Fig. 41 that a series of parallel faults results sometimes in a stepping down or up. Such faults are known as step faults.

Rift Valleys

Rift valleys are not infrequently developed within a zone of step faults, although usually the series of fault scarps is not seen because river work has filled in the lower part of the trough and thus concealed the steps. A detailed cross-section of the Rhine rift valley shows, from the evidence of well borings made in the plain of Alsace and Baden, that the dislocation is by no means due to a pair of parallel faults, but to two series of step faults which lie to either side of the axis of the rift valley (Fig. 41). The rift valley of the Jordan in Palestine shows from a relief map that a series of fault scarps is quite usual. Fig. 42 shows the inner trench, which is more than 600 feet below sea-level, with the Jordan flowing between fault scarps (Fig. 43). Nevertheless, the rift feature is apparent in other scarps set at a considerable distance from the Jordan plain and at an elevation of between 600 and 1,000 feet above sea-level. Rift valleys and their companion horsts may be found in most uplifted plateau zones of the world, including intermontane plateaus in fold-mountain areas. Outstanding rift valleys include:

1. Central Scotland, bounded by the Grampian Mountains and the Southern Uplands.

2. The Rhine Rift Valley, bounded by the Vosges Mountains and the Black Forest and their northward extensions.

3. The Torrens Rift Valley in South Australia.

4. The Jordan Valley of Palestine.

5. The great African rift valleys (Fig. 24).

The Rhenish, Scottish, and other isolated valleys of this type are relatively insignificant in comparison to the great African rifts which are far from simple structures. These rifts, which extend from Syria in the north to the Zambesi River in the south, are nearly 3,000 miles in length. The map (Fig. 45) shows the detail of these valleys. Most of the several sections of rift valley are bounded by a series of fault scarps rather than by a single scarp. Modern study has indicated that there are several points of interest which are notable in all these rifts:

1. The width of each valley is very uniform, ranging from twenty-five to fifty miles in different valleys.

2. The floors of the valleys vary very much in regard to their levels, some being well above sea-level, whilst others, such as those supporting Lake Tanganyika and Lake Nyasa, are 1,000 or 2,000 feet below sea-level.

3. The horsts which are adjacent to each valley are of unequal elevation, and tend to slope down and away from the edge of the rifts.

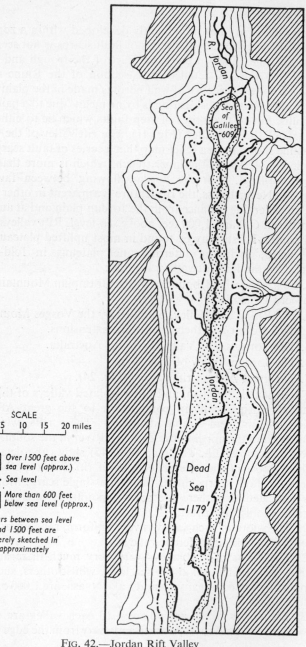

SCALE

0   5   10   15   20 miles

Over 1500 feet above
sea level (approx.)

Sea level

More than 600 feet
below sea level (approx.)

Contours between sea level
and 1500 feet are
merely sketched in
approximately

R. Jordan

Sea
of
Galilee
-609'

R. Jordan

Dead
Sea
-1179'

FIG. 42.—Jordan Rift Valley

4. Measurements of gravity in and near these rift valleys show very low values in the valleys, whilst on the plateaus gravity values are fairly normal.

Various theories, which have been developed during this century to explain rift-valley formation, discredit the possibility of a collapse of a wedge of the sial rocks between two faults. The most satisfactory recent theory is one which demands compression rather than an up-lifting force from inside the crust of the earth. It may be compared

Photocrom Co. Ltd.

FIG. 43.—Jordan Rift Valley and the Dead Sea, Palestine
In the background is a fault scarp to the west of the Dead Sea.

to the force which crushes the sediments in a geosyncline to form mountain structures, but it is operating upon a rigid sial block in this instance. It is argued that a compressional force acting along the surface will cause a rigid block to weaken and, along lines of weakness which develop, thrusting occurs as shown in Fig. 44. These thrusts result in the formation of two opposed fault scarps which override the rift block. Later, step faulting will occur on the edges of the trough so produced, whilst a certain amount of folding in the strata lying on the floor of the valley may be observed. This theory satisfies most of the modern observations on rift valleys, and in particular that of the rift block, which is virtually trapped and held down by the overthrust horsts. It is obvious that the rift block cannot rise, and until it does rise it will not be in equilibrium from the point of view of isostasy. The low-gravity observations in these rifts is indicative of a depressed mass of sial. Calculations based on the

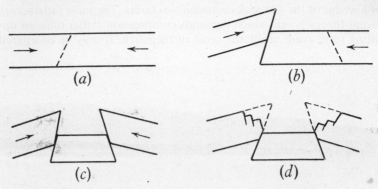

FIG. 44.—Origin of a Rift Valley through Compression

estimated thickness of the sial layer show that it would break down under great stresses to give troughs which would have the width of the African valleys. Over-thrusting would also tend to tilt the horsts away from the edges of the valleys as observed.

World Survey of the Major Earth Structures

Mountains, such as have been described as 'young', cover a considerable portion of the surface of the earth today, but there remains a much larger area which is developed on a foundation of 'older' mountains. The history of many 'old' mountains is still not fully known, but whilst they have undoubtedly originated from earlier geosynclines, more recently they have been lowered by erosion to form peneplains (see p. 76). On a peneplain, erosion by rivers is practically at a standstill, since the surface, to all intents and purposes, is at sea-level. The erosive process of running water can be renewed by the uplift of a peneplain. Thus, many 'old' mountains have been reduced and uplifted several times. In passing, it may be noted that there is evidence of many 'young' mountains having been reduced to peneplains and then uplifted.

The surface relief of many areas of the world today is that of plains or plateaus. Here the underlying structures are obscured by relatively recent deposits of rock. However, it is not entirely unreasonable to say that such lowlands, as those of Amazonia or of Northern Europe and Asia, rest on a foundation of very old rocks which probably exhibit mountain structures. Other areas may be 'old' but relatively elevated in consequence of fairly recent uplift; the mountains of Scandinavia are a good example. Such areas have been relatively stable for a long time and are, in this respect, like the shield lands of Pre-Cambrian age. Thus, the rocks of the Mississippi basin do not show that they have been disturbed to any great extent since the formation of the Appalachian Mountains, which are quite 'old' mountains.

Four distinct ages of mountain building may be related to the stratigraphical sequence (see p. 16). The earliest phase is known as Pre-Cambrian, but in the light of recent geological knowledge, it is most probable that there were several episodes of orogenesis during this very lengthy period. The two maps (see Figs. 45 and 46), which should be consulted at this stage, show that each continent is built in part of these very 'old' mountain structures. They form the regions of shield lands which encircle the Arctic basin in the northern hemisphere. They are nowhere of great elevation, but they form large zones of ancient stable rocks in the form of peneplains. They resemble shields in their shape. The 'old' mountains form the more elevated plateaus of Brazil, Africa, and Australia, together with Arabia and the Indian Deccan, whilst Antarctica is another shield. The scattered plateaus of the southern continents and Asia are regarded as the portions of an earlier and very extensive land mass which have been dispersed by the operation of continental drift.

Somewhat 'younger' mountains than the 'very old' mountains were developed in mid-Primary times, and these are known as Caledonian, since North Britain is built of mountains of this age. Fig. 27 shows the complex structure of this region in the Highlands of Scotland. There is little, if any, resemblance to the pattern of 'young' mountains in this section of the Highlands. Again, all continents show some evidence of this period of mountain construction. Most of the north-western part of Europe is Caledonian, whilst, to the east of Norway, lies the shield land of Pre-Cambrian rocks. To the south of these two zones lies the third, or Hercynian, mountain region of Southern Ireland, South Wales, and Southern England, with much of Western and Central Europe. The Hercynian mountains (alternatively, Armorican or Variscan) are late Primary in age. There are no very elevated mountains of this age in Europe, although some areas have been uplifted to form plateaus, such as the Central Plateau of France and the Bohemian block. In North America the Appalachian Mountains show evidence of both Caledonian and Hercynian developments.

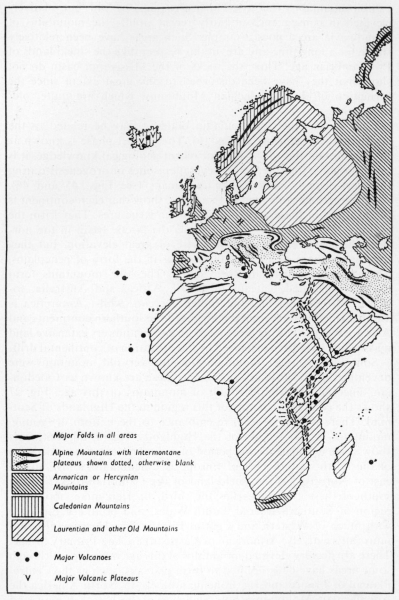

FIG. 45.—Structure

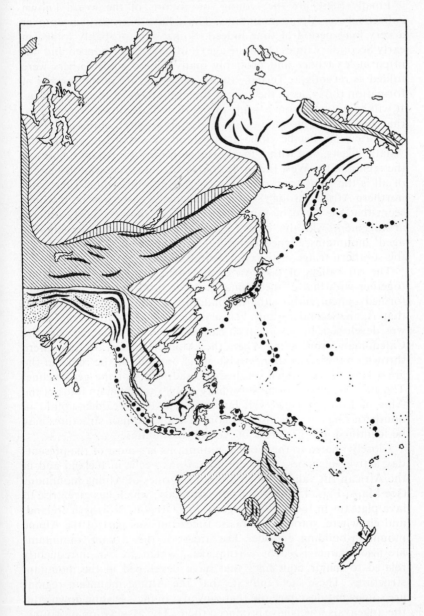

of the World—I

Finally there are the 'young' mountains of the world known collectively as Alpine mountains. These have been thrown up over a very long period of time indeed. Some were probably folded in early Secondary times (Jurassic age) and have been reduced and up-lifted at least once and probably many times; whilst others were folded as recently as Tertiary times, and some are in the process of formation today, as in the case of the mountains of the East Indies. It will be noticed, from a study of the two maps already referred to, that the Alpine mountains are marginal to the continents. They form a circum-Pacific girdle, with island mountains in arcs on the Asiatic coasts, and with a local extension into the Atlantic basin in the region of the West Indies. Probably the most remarkable feature of all is the great train of east–west Alpine folds which runs from northern Africa, through southern Europe, to eastern Asia. It would seem that the build of Euro-Asia, at least, shows a fairly orderly arrangement, with the 'oldest' mountains in the north, the 'middle-aged' mountains aligned centrally, and the 'young' mountains on the southern fringe of the continental mass.

The rift valleys of East Africa and their extensions in Palestine, together with that of the Rhine, are structural features which were formed concurrently with the building of the Alpine mountains, that is, in Secondary and Tertiary times. The Scottish rift valley was developed, however, much earlier and was associated with the Caledonian orogenesis. When the Alpine-Himalayan folds were thrown up there was an inevitable effect on some of the rocks in the great land blocks which crushed the sediments of the geosyncline. The Jura Mountains were subsidiary ripples to the main folds of the Alps of Europe, whilst still farther north the Wealden upfold in Southern England was also developed. Similar minor structures may be identified in other areas of 'young' mountains.

Closely related to the 'young' mountains are most of the present-day active volcanoes, although exceptions occur in Iceland and in the African rift valleys which are not in zones of Alpine mountains (see Maps, Figs. 45 and 46). The vulcanicity, which has given rise to lava plateaus in India, Washington and Oregon, Northern Ireland, and elsewhere, started in Jurassic times but was part of the Alpine mountain-building episode. The areas of the 'young' mountains are also characterized by earthquakes, which are not infrequently related to major fault lines that have developed in the mountain structures. These facts indicate that the Alpine mountain regions are somewhat unstable, and that they are, in fact, settling down after the upheavals which have produced them. The association of 'young' fold mountains with vulcanicity and seismic disturbances is a fairly

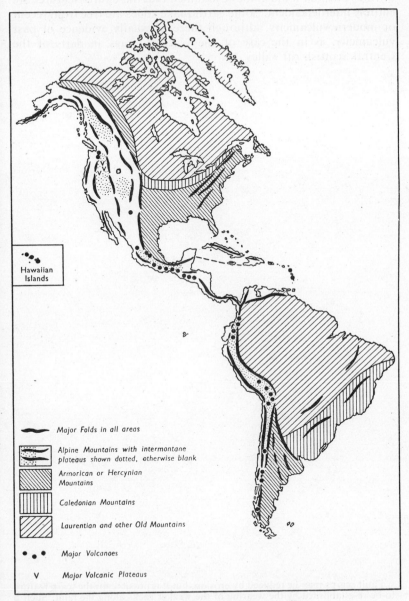

Hawaiian
Islands

Major Folds in all areas

Alpine Mountains with intermontane
plateaus shown dotted, otherwise blank

Armorican or Hercynian
Mountains

Caledonian Mountains

Laurentian and other Old Mountains

Major Volcanoes

V   Major Volcanic Plateaus

Fig. 46.—Structure of the World—II

obvious one, but it must not be assumed that these three features are entirely interdependent. Some earthquake zones are free from recent or modern vulcanicity, although there is usually evidence of past vulcanicity, as in the case of the volcanic crags in parts of the Central Scottish rift valley.

Fault scarps may be reduced by erosion, but it is possible for the scarp feature to be re-established by the wearing of the weaker rocks along the fault. Such a scarp may be almost indistinguishable from a fault scarp but, as it is an erosional feature, it is better described as a fault-line scarp.

Part II

EARTH SCULPTURE

CHAPTER V

THE WORK OF RIVERS

Introduction

In a previous chapter, various processes by which the surface of the rocks is weathered were discussed. Weathering in itself is a static process, and may be regarded as an essential preliminary to the dynamic action of erosion or denudation. Once weathering has produced small fragments of rock, then it is the turn of other agents to remove these by transportation to other positions. Although ice, wind, and the sea are powerful erosive agents, it is unquestionable that a very large proportion of the area of the land surfaces is eroded by the action of running water, and thus rivers are most important tools in the work of earth sculpture.

It has been indicated how rock surfaces may be disintegrated, and how the resultant particles are transported by the water of rivers for long or short distances (p. 8). The finest of these particles have the sea as their ultimate destination. Everywhere running water plays its part in this movement, apart from some exceptional areas, and the various transported fragments, which constitute the load of a river, are used for further erosion of rock. In most areas of the world today is seen the effect of river work, and this has persisted for at least twenty to twenty-five thousand years. In parts, other than those which were covered with extensive ice sheets in the Pleistocene period, river work has been constantly in operation for a much longer time. Thus the initial stages of river development are not readily seen today, since the majority of rivers have reached a fairly advanced stage in their evolution.

Graded Profiles

An examination of the course of a river will show that it is capable of subdivision into sections, each of which displays fairly well-defined characteristics. A profile section along the length of the river will often permit such tracts or subdivisions to be distinguished by the changes in the gradient of the river-bed. On the whole the longitudinal profiles of many rivers approximate to an 'ideal' curve, as shown by Fig. 47. The gradient of the river usually decreases downstream, so that the river is swift in its upper course and much slower in its lower sections, but it must be noted that towards the sea it has an increasing volume furnished by its tributary streams. Cross-sections, which may be related to the longitudinal profile given in

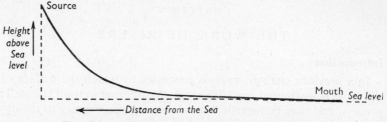

FIG. 47.—Longitudinal Profile of a River Course

Fig. 47, show that the valley of the river tends to widen in a down-stream direction, although preserving a characteristic V-shape (Fig. 48).

The approximation of many longitudinal profiles to that of Fig. 47 arises from the fact that all streams are attempting to reach a state that is known as graded. It must be remembered that rivers are collecting weathered material from the surfaces over which they flow, and this material is carried along by the water in solution and in suspension, whilst large fragments are rolled along on the bed of the river. Such is the load of a stream (Fig. 49). The undissolved fragments are capable of reducing other rocks against which they are projected by the stream flow, and so further erosion occurs. Thus the load of a river may be increased, and it is possible that the load becomes too great for the river to move, and unless the speed of the river is increased, deposition must take place. With a graded profile it may be said that the river is in equilibrium throughout its course, in that its speed and volume at any point are just sufficient to carry all the load and eventually deposit it in the sea. Such an ideal state of affairs is rare, if not non-existent.

FIG. 48.—Progressive widening of the Valley

Let it be assumed that the course of a river is developed on a land surface which was initially irregular at the time when streams began to flow on it (Fig. 50 (a)). Clearly the gradients are very varied along the course of such a river. The maximum load is likely to be at the lower end of the course, and it is here that the profile becomes

J. Hardman

FIG. 49.—The Dry Bed of a Stream in Langdale, Westmorland
This shows some of the load carried by a river.

rapidly graded in relation to the level of the sea. In these early stages the river is concerned with cutting down, or degrading, its original profile, so that it assumes the appearance shown in Fig. 50 (*b*). Three changes in the profile have occurred. Firstly, there has been a general lowering of the surface from the original XY to ZAY. Secondly, the lower part of the course is graded from A to Y. Thirdly, there has been an extension of the head region of the course to Z. Normally progressive development of the lower-graded section will proceed in

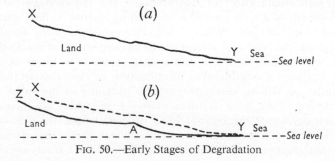

FIG. 50.—Early Stages of Degradation

an upstream direction, so that an approach is made to the graded profile of Fig. 47.

Degradation and Aggradation

Such a graded course would be easily disturbed by alterations in the flow of the river. For instance, a temporary influx of storm waters in the upper course may result in an increase of down-cutting which will flatten locally the graded curve, and yet this same down-cutting will increase the river load and there will be deposition of rock fragments at some point farther downstream, where the load is too great for the speed of the river. At such points of deposition or aggradation there will be a local steepening of the graded curve which in turn will be degraded. Frequently in the early development of a river's course there is some feature which subdivides the course into two or more sections, each of which is practically graded. A band of resistant rock may lie across the course and produce such a division, marked by a waterfall or rapids in the river-bed (Fig. 51). The

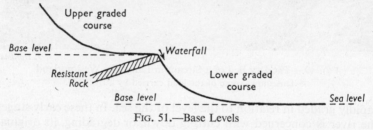

FIG. 51.—Base Levels

removal of the fall in the course of time will result in the merging of the two curves, so that one continuous curve is produced. In Fig. 51 horizontal lines are drawn which are called base levels. Every graded curve is related to such a level, which is a horizontal plane passing through its lowest point. The base level of a completely graded stream coincides with sea-level, as has been mentioned.

The immediate effect of degrading in the initial stages of the development of a river course is to produce narrow valleys with V-shaped cross-sections. Whilst such cross-sections were common throughout the course of a river in the earliest stages of its development, with the establishment of a graded profile in its lower sections there has been a considerable modification of the cross-section of the valley, as will be seen later, with a widening of the valley and a flattening of its floor to form plains. Even where river courses are well developed throughout most of their length there is some section, usually at the head-stream area, which shows the characteristics that are typical of this early development. Such tracts are often called youthful, but this is a somewhat misleading description since most of the river course must be regarded as of the same age. It is more correct to regard such sections with their narrow, V-shaped valleys as not having reached the advanced development shown elsewhere in the course. Obviously such less mature features may be produced anywhere in the course where resistant rocks retard the opening up of the valley or where some other factor is at work, such as rejuvenation (p. 85).

Immature Valleys

The chief features of such relatively undeveloped tracts in the course of a river are as follows. The stream has a fairly small volume of water but a fairly high speed, because of the considerable gradient, and cuts a valley which is steep-sided with a deep, V-shaped cross-section. Original bends in the stream, due to local irregularities, result in the formation of overlapping or interlocking spurs on the valley sides. These spurs make it difficult or impossible to see far up or down the river course at stream level (Fig. 52). The bed of the

The Aircraft Operating Co. of Africa (Pty.) Ltd.

FIG. 52.—Kafue River Valley, Northern Rhodesia

An immature valley showing V-shaped cross-section and interlocking spurs. The valley slopes are covered with dry forest.

stream is littered with boulders and large fragments of rock which must be reduced in size by the mechanical action of the running water before they can be removed downstream (such boulders may be moved short distances by storm waters when the momentum of the river is much increased). Resistant rocks produce waterfalls, marking irregularities in the longitudinal profile, and these must be reduced before gradation is accomplished. Elsewhere, such resistant rocks may be reduced by pot-holing. Side-streams form gullies which may develop into more open amphitheatres separated by the spurs of the valley sides. The contour diagram in Fig. 53 shows a section of a river course in this immature stage. The average gradient is about 1 in 10, but at A the close spacing of the contours indicates that for a few yards the gradient is very much greater, marking a waterfall. To some extent the winding river is seen to fit into a series of interlocking spurs.

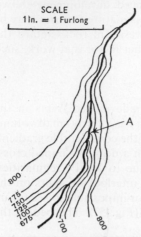

SCALE
1 in. = 1 Furlong

FIG. 53.—Upper Valley Tract
(Kingsdale, near Ingleton)
(Contour values in feet)
(*Based upon the Ordnance Survey Map with the sanction of the Controller of H.M. Stationery Office*)

Pot-holes are developed by the swirling action of loose rocks on the stream bed, with the result that a hole is ground into resistant rock. This hole is deepened by the drilling action of the boulder as it mills round in the hole that it has made. This action smooths the drilling boulder so that eventually it resembles a cannon-ball. A series of pot-holes may unite to form a water-course which is like a slot in the land surface whose width may be very much less than its depth (Fig. 54). The Strid on the River Wharfe in Yorkshire is an example of a slot-like ravine of this type.

FIG. 54.—
A Ravine

Waterfalls and Lakes

Waterfalls are very common in the ungraded sections of a river course. The life of such falls is short, since the stream proceeds to remove these obstructions with all speed by down-cutting. Waterfalls are formed by layers of resistant rock which have yet to be graded in relation to the adjacent rocks (Fig. 55). In some cases the fall is determined by a horizontal layer, which acts as a protective capping to less-resistant layers of rocks, as shown in Fig. 56. It is possible for a fall to develop at a later stage in the history of the

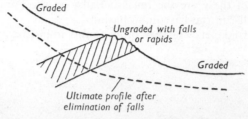

Fig. 55.—A Waterfall

river, as for instance where a sill of igneous rock which was originally below the surface of the land is reached and exposed by the downcutting of the stream. Falls recede in an upstream direction, and in the case of the Niagara Falls the recession has produced a gorge

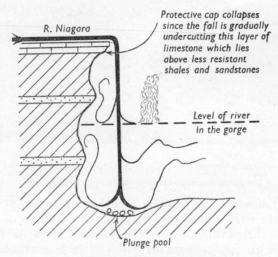

Fig. 56.—Niagara Falls

on the downstream side of the falls (Fig. 57). This state of affairs will persist until the fall is completely reduced and a graded stream will then take its place.[1] Even more spectacular than the Niagara

---

[1] The recession is now at the rate of about 2½ feet per annum, and the computed time for the complete recession of the falls is at least 2,000 years.

gorge is that on the Zambesi below the Victoria Falls. Some of the highest falls are located in South America, where the ancient and resistant rocks which form the Brazilian Plateau and the Guiana Highlands interrupt the courses of streams as they leave these uplands. Two of these are the Iguassu Falls, on the Parana, and the Kaieteur Falls, in the Essequibo basin.

Lakes and lakelets are very often found immediately above the

The Niagara Falls Power Co

FIG. 57.—Niagara Gorge and Falls, New York, United States of America
The gorge has been cut with the recession of the falls.

site of a fall, since the resistant rock layer also ponds back the water of the river. Lakes are also temporary features of the landscape, and their existence may be terminated in one of two ways. Firstly, the barrier damming the waters of the lake may be reduced so that the lake is drained. Secondly, the still waters of the lake will check the speed of the river as it enters the lake on the upstream side, resulting in a settlement of part of the load of the river, and eventually the lake will be filled in and thus eliminated. Sometimes deposits from side streams may subdivide a lake.

Valley Widening

Throughout its course in its early development a river is presumed to have cut down as rapidly as possible in an attempt to attain a state of gradation. A glance at the graded profile in Fig. 47 will show that a river cannot cut down into the land surface below its base level (this is, in general, sea-level). Long before the curve of the longitudinal profile has been lowered to the position indicated in Fig. 47, the down-cutting process will have been replaced, in part, by side cutting. Most probably with the relatively advanced grading in the lower sections of its course, as compared with the still ungraded upper course, the load received in some parts may prove to be too great for the river to carry, since its speed is considerably reduced. Thus deposition or aggradation is likely to accompany the

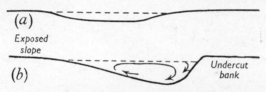

FIG. 58.—Change of Cross-section of Stream Bed with Side Cutting

onset of side cutting. Some examination of these processes is necessary, since they are of great help in the widening of the valley and for the planation of its floor.

Initially the stream, as well as its tributaries for that matter, is rarely straight, and at all the bends, associated with interlocking spurs, side cutting is likely to commence. This side cutting results in a change in the cross-section of the stream-bed (Fig. 58). Apart from lateral currents, there are also vertical movements in the stream which hasten the widening and deepening of the minor bends into meanders (see Fig. 58 (b)). This produces a distinct shift of the stream course, and visual evidence of this is afforded by the gentle, crescentic slopes that are exposed on the inner banks of the meander curves, whilst on their outer banks there are steep, undercut slopes. The gentle, crescentic

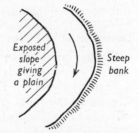

FIG. 59.—The Beginning of a River Plain

slopes are the beginnings of a river plain (Figs. 59 and 60), and they may be built up by storm deposits after the river has been in full

Aerofilms Ltd.

FIG. 60.—The Beginning of a River Plain, Wye Valley, Monmouthshire

Note the crescentic areas in the foreground where spurs have been reduced with the widening of the valley.

spate. These features are to be seen in the majority of river valleys and, on miniature scale, in brooks of quite modest proportions.

Development of Plains

Further development of this plain is effected by the shift, downstream, of the meanders. This movement affects the spurs, which were interlocking but now are worn back on their upstream sides. Eventually the spurs are set back in line with the valley sides. The valley is relatively open, and visibility upstream and down is much improved. This sequence of events is shown in Fig. 61, and in the last stage it will be seen how large is the area of the plain. The development of the river plain by side cutting is also shown in Fig. 62, which indicates the much-reduced gradient of the stream, since it falls only about thirty feet in approximately six miles in the course on the map. Irregularities in the longitudinal profile may still exist in the form of minor rapids, but, largely, the stream is graded.

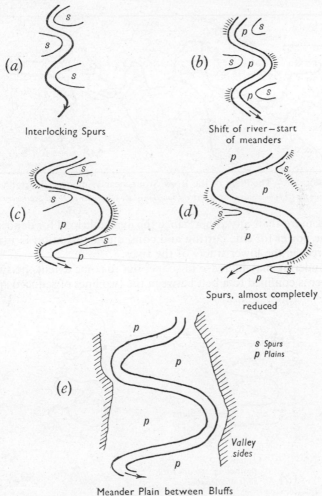

(a)

Interlocking Spurs

(b)

Shift of river – start of meanders

(c)

(d)

Spurs, almost completely reduced

s Spurs
p Plains

(e)

Valley sides

Meander Plain between Bluffs

Fig. 61.—Development of Meanders and Plain

Such plains may be found almost anywhere in the course of a river, with the probable exception of the headstream zone. If a river course is subdivided as shown in Fig. 51, having at least two tracts with independent base levels, then an upper and lower plain are likely to be formed. With the acquisition of more and more tributaries and therefore an increase in deposition, these plains will widen still further. The plain which is nearest the sea is likely to increase

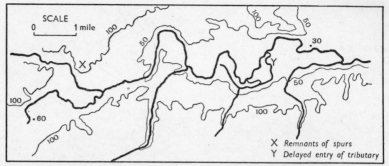

FIG. 62.—The Plain of the River Stour, near Wimborne, Dorset
(Based upon the Ordnance Survey Map with the sanction of the Controller of H.M. Stationery Office)

in area more than any other, since the available load for deposition and the power for side cutting and consequent planation is usually very great in the lower tracts of the river.

Meandering may increase in scope, but to some extent the swinging river is confined to a belt between the two lines of reduced spurs,

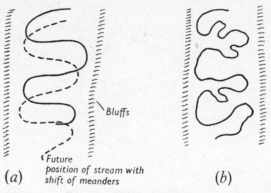

FIG. 63.—Shifting Meanders

and further side cutting leads to the formation of highly involved and contorted meander curves (Fig. 63). It is clear that this process has lengthened very much the course of the river, but the stream is very sluggish. At the same time the migration of the meanders planes very thoroughly the lowland over which the stream is flowing.

The river may continue to widen its plain in this fashion, but, as the valley sides of the main stream and its feeders are completely reduced, these several plains merge into one very extensive, lowland area. The relief of these ultimate plains is so slight on either side of the rivers that an increase in the volume of the water, consequent upon heavy rains or perhaps a sudden melt of snow, will inevitably lead to their flooding. Hence this lowland zone is frequently known as the flood plain (Fig. 64).

Geological Survey photograph reproduced with the permission of the Controller of H.M. Stationery Office

FIG. 64.—Strath Glass, Inverness-shire

Meanders of the River Glass in a wide, alluvium-filled basin. The valley is developed along a line of faulting.

Meandering lower courses of large rivers are easily identified on atlas maps, and it is of interest to note that the name 'meander' is derived from Meanderes, a river of Asia Minor. Amongst other rivers with pronounced meandering courses are the Mississippi, the Forth, and the Seine. An allied feature, developed from meanders, is the oxbow, a type of lake. Oxbows are formed when a meander loop is breached by storm waters and then sealed off by the deposition of the excess load which is carried at such times by the river (Fig. 65). An oxbow is a form of lake which, in general, has but slight importance. An alternative name is 'mortlake', and a suburb of London near the meandering Thames has this name, since it lies near a drained oxbow. Deposition consequent upon the sluggishness of the stream may cause it to build up its bed, thus confining itself

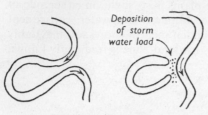

FIG. 65.—An Oxbow

between natural levees of mud and silt (Fig. 66). The Po, the Hwang-ho, and the Mississippi afford good examples of this process at work in their plains. A breach of a levee is disastrous, since flooding is very extensive over the plain. The natural levees may be artificially reinforced to obviate such floods as have swept the Mississippi lowlands in certain years of this century, although local success has come by tackling the problem from outside the flood plain region in the upper basin of the Tennessee River, an important tributary.

The natural levees may divert tributary streams so that they flow parallel to the main stream until some local advantage permits an entry into it (Fig. 62). There may also be formed a braided course through the choking of the water channel by a dropping of a considerable part of the river's load which produces a bifurcation of the

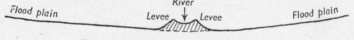

FIG. 66.—The Levees of the Flood Plain

stream (see Fig. 82). This process may become widespread, and there are quite a number of local names for these several branches of a braided stream. Anabranch and billabong (from Australia) are two terms applied to such branch streams.

Estuaries and Deltas

A river enters the sea at the end of its course through an estuary or a delta, the latter often being a seaward extension of the lower plains. It has an estuary or a mouth if tidal scour is strong enough to assist in clearing away the load which is dropped abundantly in this part of the course. However, salt water helps to precipitate much of the load, as it causes the finest particles to coagulate. The estuary is much wider than the true river, and is filled with sea water at high tide. At low tide the river swings into the sea through extensive banks of mud and sand which are exposed at that time, and, in fact, the course is much braided. If, on the other hand, the sea into which the river is flowing has weak tides, then a delta is usually formed. The lack of tidal scour does not adequately account for the formation of a delta, since some rivers entering tideless seas have no deltas, whilst some deltas are built on the shores of seas with moderately large tides. Two other factors must be taken into consideration for the explanation of full-scale delta formation. Firstly, the river must have an upper course which is far from graded, and

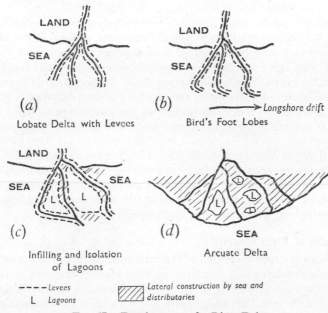

(a) Lobate Delta with Levees

(b) Bird's Foot Lobes ⟶ Longshore drift

(c) Infilling and Isolation of Lagoons

(d) Arcuate Delta

- - - - Levees
L   Lagoons

Lateral construction by sea and distributaries

FIG. 67.—Development of a River Delta

therefore capable of yielding a large load. Secondly, this load must be deposited on a grand scale in the lower plain tract. Choking of the river channel will lead to braiding and the formation of distributaries near the sea. Distributaries are distinguished from tributaries in that they lead water away from the main stream. Each distributary disposes of its load by extending the natural levees into the sea, so that the initial form of most deltas is lobate (Fig. 67 (a)). The delta of the River Nile, which was the classic example of the Greeks, represents a further development from the lobate type. The natural levees projecting seawards along the sides of the distributaries are converted into curved spits of silt according to the prevailing drift in the sea immediately offshore (Fig. 67 (b)). These spits are eventually linked with one another and enclose areas of water called lagoons (Fig. 67 (c)). The Rhône delta shows this stage of development today. Lagoons have a limited existence, since their infilling is a matter of time. Lastly, the area of the delta is increased laterally by drift and current deposition as well as by the work of minor distributaries, thus creating the more typical arcuate outline as exemplified by the Nile delta (Fig. 67 (d)). The resemblance of this outline to the Greek letter Δ (delta) has given the term which is so widely accepted.

Local sea action may produce somewhat unusual outlines to deltas, but the essential features will be present. Atlases of the nineteenth century may be compared with modern editions, and it will be seen that there has been a considerable change in the last fifty years in the outline of the Mississippi delta. An outstanding example of the diversion of a stream by the blocking of distributaries is given by the Hwang-ho. Since the middle of the nineteenth century this river has entered the sea both to the north and south of Shantung at various times.

Small deltas may be produced by side streams entering lakes. In some instances such lake deltas, extending from opposite shores, may eventually subdivide a lake. In Cumberland, Buttermere and Crummock Water have been formed in this way, and in Switzerland, Brienzersee and Thunersee are separated at Interlaken.

Examples of River Courses

Students are recommended to attempt an analysis of river courses, based on the foregoing account, by reference to maps of various scales. The map given in Fig. 68 is of the River Esk in Cumberland, and it will be seen that this stream has a valley which widens progressively towards the sea. In the twelve miles of its course to the head of its estuary, the river descends from over 1,000 feet to under 400 feet above sea-level in about two miles, and the lower half of its course is a meandering one, across a fairly wide plain below 200 feet. It is fairly clear that the upper course is a narrow valley, where the river is probably incompletely graded, which ends roughly at the 400-foot contour. High land lies on either side of the plain near the sea, and this has yet to be reduced by the Esk and its tributaries before this plain is completely developed.

A relief map of the British Isles in an advanced atlas will show the salient features of the course of the River Severn, which is one of considerable interest. Not far from its source area in Plynlimmon the river enters an upland plain between Llanidloes and Newtown, centred on Caersws, but below Newtown it flows through a narrow valley almost as far as Welshpool. East of the latter town is the wide plain of the Camlad, a tributary stream. Below Welshpool the main river enters a very extensive flood plain, across which it meanders until it has passed Shrewsbury. From Buildwas to beyond Bridgnorth it flows once again in a gorge-like valley with steep tributary valleys entering from either side. Below Stourport the Severn enters

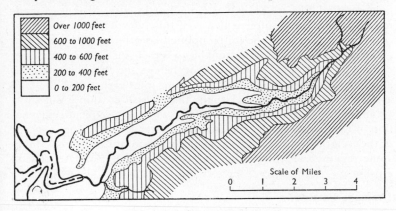

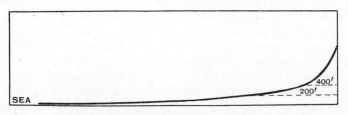

.Fig. 68.—River Esk, Cumberland

Relief Map of the Valley (*above*)
Approximate Profile of the Valley (*below*)

(*Based upon the Ordnance Survey Map with the sanction of the Controller of H.M. Stationery Office*)

its lower plain leading to the estuary. It is perhaps unfair to cite this example without brief comment. The Severn is a river that has suffered diversion by glacial damming, which helped to produce the upper plains and the intervening gorges. It is not desirable to amplify this statement here, but further discussion of the Severn will be found on p. 110. However, it is a good example of a river which has a valley that is successively wide and narrow. Other causes than glacial damming may produce similar variations in the width of the valley throughout the course of a river, and some of these are discussed in another part of this chapter.

The Cycle of Erosion

The work of rivers will eventually lower the land surface to an almost level plain not very much above the level of the sea. There must be, even at this stage, some slight slopes to enable rivers to maintain their flow. The final form of the land surface is known as a peneplain, but there will be a few residual hills of low elevation which will represent remnants of the land surface that has been reduced. Such residual hills are known as monadnocks, after Mount Monadnock, in New Hampshire, which is of this type. The complete process of reduction from the initial uplifted land surface is known as a cycle of erosion. It is questionable whether such a cycle is ever completed. It would seem that long before the time required for absolute peneplanation has elapsed, some interference with the land surface will have occurred. Usually the cycle of erosion is cut short by an uplift of the eroded land surface, with a consequent rejuvenation of the rivers, as will be explained later (p. 86). The concept of the cycle of erosion may be applied to erosion of arid areas, limestone regions (where special characteristics arise), and to coastal erosion.

Drainage Patterns

A study of fairly detailed maps will show that a pattern is discernible in the streams of most river basins. A river basin is an area which is drained by one complete set of streams whose waters reach the sea by a common main stream. The basin is enclosed by a line that follows the crest line of the highland margins which exist. This is the main watershed of the basin. It is fairly true to say the pattern of the streams within a basin will be one of two types. These two patterns are called dendritic and trellised, the latter being sometimes referred to as gridiron. The former type is characterized by a series of streams which resemble the branches of a tree as they converge to its trunk (Fig. 69, also Fig. 22). With the second or trellised pattern, the main stream is usually well defined throughout the basin, and its feeders join one another and eventually enter the main stream approximately at right angles to it. The resultant pattern is thus more regular in appearance than that of the dendritic type (Fig. 70, also Fig. 31). The two patterns are produced by different circumstances. In the dendritic type the direction of river flow is controlled by downgrade slopes, whereas in the second case a very considerable

FIG. 69.—Dendritic Drainage Pattern

measure of control is exercised by geological structures. Rarely does a river basin exhibit one of these patterns throughout, and not infrequently a mixture of the two may be observed. If the whole basin is worked out in the same type of rock or similar rocks, then the dendritic pattern will almost certainly develop, particularly if the rocks offer uniform resistance to denudation by running water. If, on the other hand, the basin has different rocks which are of varying resistance, then the chance of the development of trellised drainage is considerable.

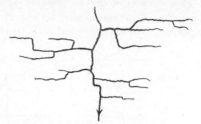

FIG. 70.—Trellised or Grid-iron Drainage Pattern

Trellised Drainage and Escarpments

Trellising is most simply developed as shown in Fig. 71, where in (*a*) an initial stream which is flowing fairly directly towards the sea crosses alternate outcrops of weak and resistant rocks. At the resistant bands, side streams are largely absent and the valley is narrow, whilst upsteam and down from this point the development of side streams is evidently easy along the bands of less resistant rock. These side streams tend to flow parallel to the edge of the resistant rock (Fig. 71 (*b*)). Obviously all the side streams continue to cut downwards in order to grade their respective valleys, but whereas they have lowered the level of the less resistant rocks over a wide area, the more resistant rock bands will not be so reduced except at the points where the main streams cross them. This results in the isolation of an upstanding band of resistant rock known as an escarpment, which has a relatively deep, narrow, and steep-sided gap cut into it at the point where the main stream passes. The development of escarpments and their river gaps is shown in Fig. 71 (*c*). Examples of such river gaps are innumerable. The relation of the more gentle slope of the escarpment to the dip of the resistant rocks is reflected in the name of dip slope which is given to it. The steeper slope is called the scarp slope or edge (see Fig. 71 (*d*) and Fig. 144). Escarpment development is a local feature, although lines of scarplands may persist for several hundreds of miles where a resistant rock band is so disposed. Whilst an escarpment may appear to be a very significant element in the landscape, it is destined to be reduced. In Fig. 71 (*c*) it will be seen that the feeders of the main stream have acquired their own tributaries, some of which descend the steep scarp slope. These streams must reduce this slope, and eventually a recession of the whole scarp occurs. This recession leaves greater space for the development of side streams flowing parallel to the chief feeders, and thus the trellised pattern will be extended still further. The limestone escarpment of

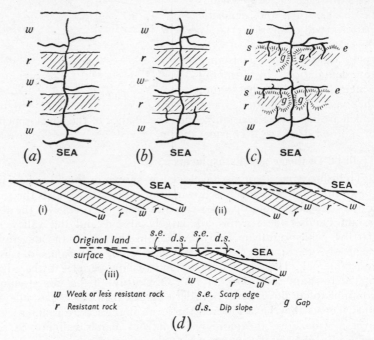

FIG. 71.—Development of Trellised Drainage with Escarpments and Gaps

Note (i), (ii), and (iii) in (*d*) refer to north-south sections in (*a*), (*b*), and (*c*) respectively.

the Cotswold Hills shows that recession has occurred near Winch-combe, but there remain isolated parts of the previous scarp front in Bredon Hill and its neighbouring summits (Fig. 72).

Trellising is also a possible consequence of drainage in folded structures where parallel streams may flow in the downfold valleys, and at certain weak places in the upfolds gaps may be formed with connecting streams at right angles to the general trend of the drainage. The drainage of south-western Ireland is controlled in this fashion, as shown by the map given in Fig. 139. This pattern may also arise where faults or joints control the drainage lines. Two areas which show trellised drainage very well are the Wealden

area of south-eastern England and much of the Appalachian valley
region of the United States of America (see Fig. 31, p. 37).

The development of trellised drainage has called into being a
nomenclature for the various types of stream. The initial drainage

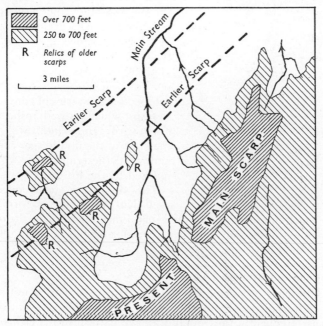

FIG. 72.—Recession of an Escarpment—Cotswold Hills between
Cheltenham and Broadway

(*Based on the Ordnance Survey Map with the sanction of the Controller of
H.M. Stationery Office*)

in consequence of the general slope of the land surface towards the
sea produces consequent streams (C) as shown in Fig. 73, whilst the
later tributaries, more or less at right angles to these streams, are
called subsequent streams (S). The rôle of these streams is clearly a
major one in escarpment formation. Later still the pattern is enriched by more streams, which are, in general, short and act as
feeders to the subsequent streams. These are relatively minor
streams, and tend to flow at right angles to the subsequent rivers.
Some are known as minor consequent streams (MC) and others as
obsequent streams (O). These flow therefore nearly parallel to the
original consequent rivers, the former like the consequent rivers in a
seaward direction, although they never reach the sea as independent

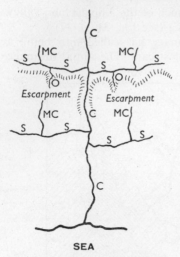

FIG. 73.—Consequent, Subsequent, and Obsequent Streams

streams, whereas the latter are opposed to this direction in their flow. The rôle of the obsequent rivers is a fairly important one, since they are largely responsible for the reduction and recession of escarpments. The drainage pattern of the Weald shows all four of these types of river. The original upfold of the Weald has given rise to two sets of streams each having developed from a series of consequent streams which flowed respectively northwards and southwards from the crest of the upfold that ran east and west, the remnant of which is marked by the Forest Ridge today.

The discussion has concerned itself with escarpments formed in rocks which dip seawards. If the general dip is landwards the scarps will face seawards and then their recession would be accomplished by the minor consequent streams.

Radial and Superimposed Drainage

The drainage of some areas is radial (Fig. 81). This is not necessarily entirely distinct from the more widespread dendritic or trellised patterns; frequently this radial drainage pattern shows elements of both the other types. The chief characteristic of radial drainage is largely dependent on the structural features of a dome. Usually upfolding produces such an upland region which is comparable to a domed roof or cupola, in that its surface slopes downwards in all directions from the summit. Another cause may be the injection into the crustal layers of the earth of a mass of igneous rock which has forced the surface up. Erosion may remove some of the surface layers of the dome and expose the underlying rock, as in the case of the granite boss which forms the Dartmoor region today. Drainage on such domes tends to be divergent in all directions from its higher levels. Perhaps the most perfect example of this type of drainage is to be seen on a volcanic cone which is only slightly denuded (Fig. 22). In Central France, in the region of the Cantal (a much dissected volcanic cone), the radial drainage has isolated a series of triangular plateaus known as planèzes. Cotton, the New Zealand geomorphologist, has suggested that these may be regarded as a distinct type of land form. The radial pattern is often developed on a surface which consists of homogeneous rock, although in the Lake District of England, where this type of drainage is found, the streams are flowing over rocks of various kinds. It is considered by geologists that the area was formed as a dome with a surface of Secondary rocks that covered both igneous and Primary rocks. On the former surface an almost symmetrical pattern of radial drainage must have been established, but once the streams had removed the whole of this

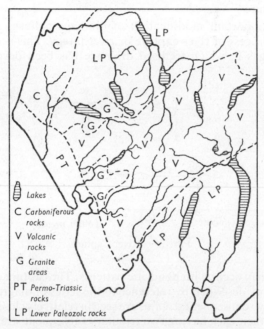

FIG. 74.—Superimposed Drainage of the Lake District

(Based on the Ordnance Survey Map with the sanction of the Controller of H.M. Stationery Office)

layer, they cut into the underlying rocks and maintained the pattern which was thus superimposed upon these older rocks. Minor modifications have occurred where tectonic disturbances have produced marked faults in the older rocks and local control of the superimposed streams has followed. Thus in places it is possible to discern a rectilinear pattern, as may be seen on the map, Fig. 74.

Superimposed drainage is not necessarily a subsequent development of radial drainage. In the area of the South Wales coalfield the eastern valleys show a remarkable alignment from north to south.

The drainage, as in the case of the Lake District, was initiated on a surface of younger rocks which have now disappeared, but the drainage lines have persisted despite the general east–west downfold axis of the coal basin, which might have produced a main axial stream flowing parallel to its trend. However, the earlier down-cutting has been strong enough to overcome the structural controls of the downfold on which the relatively simple north–south pattern of drainage has been superimposed.

River Capture

A very important modification of drainage patterns may result from the process of river capture or river piracy. This is quite common in areas in which trellised drainage has been developed, al-

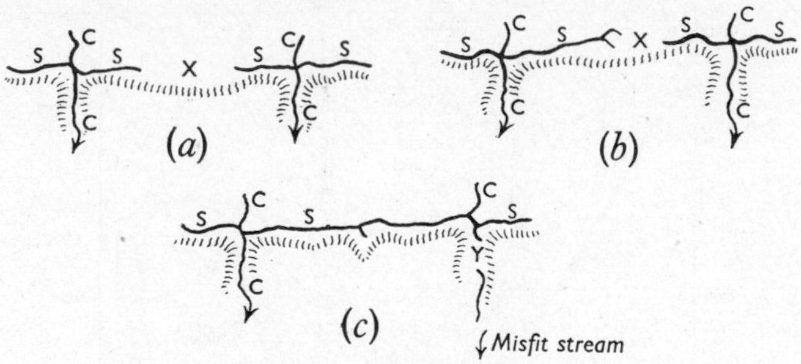

FIG. 75.—River Capture

though it may occur with dendritic patterns. The capture of a stream is most easily described by reference to the series of diagrams in Fig. 75. In (a) two consequent streams are seen in relatively close proximity to one another, each having two tributary subsequent streams whose heads are separated by some distance at X. Now it is possible for a stream to increase its drainage area by the development of minor feeders in the headstream region. This process of extension of the drainage may add to a dendritic pattern in some areas. These additional minor feeders increase the erosive power of one of the subsequent streams, and it encroaches upon the other stream. Thus in (b) the distance at X is reduced. It should be realized that one stream must have greater erosive power, and will therefore have lowered its bed in relation to that of the adjacent shorter subsequent stream. Thus, in (c) the more powerful stream has reached the head waters of the other stream, and has lowered the head region of this stream sufficiently to divert its flow. It must be emphasized that throughout this process all the rivers have been cutting down continuously, but that this diversion depends upon the much greater opportunity which one river system has compared with the other. Immediately after capture has taken place, the successful stream becomes even more powerful through the addition of water derived from the other river. It therefore lowers its bed very rapidly, so that at Y in Fig. 75 (c) not only does a gap occur, but in a relatively short time this gap is standing at a higher level than the bed of the streams which have been diverted by capture. This gap is usually without drainage and is called 'a dry gap' (Fig 76). Often there is a reversal of stream flow when capture occurs, and at Y (Fig. 75 (c)) there arises

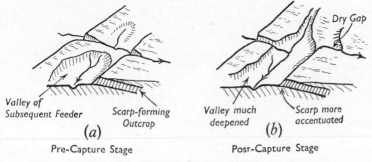

Valley of
Subsequent Feeder   Scarp-forming
                       Outcrop
         (a)

Pre-Capture Stage

Valley much   Scarp more
deepened   accentuated
        (b)

Post-Capture Stage

Fig. 76.—Formation of Dry Gap by River Capture
(*After Davis*)

what is essentially an obsequent stream, although of different origin from those referred to above (p. 79), since it arises from a reversal of drainage following capture.

Gaps in Escarpments

The lower part of the drainage basin of the captured stream is arrested in its further development, so that it may have relatively steep slopes on its sides (representing a less mature valley), although rapid and forceful erosion is well nigh impossible, since it can only draw upon a very limited surface area for its supply of water. In fact, the river is receiving so little water in comparison to what it had at its disposal in pre-capture days that it appears to be too small for its valley, and so it is called a 'misfit'. In some escarpment areas which have been formed in limestone and chalk the perviousness of the rock permits this meagre misfit stream to sink underground, and so there is literally no surface drainage and the valley becomes 'dry'. Only when the water-table in the rocks is elevated temporarily by seasonal rains do surface streams appear. The Winterbourne villages in the various dry valleys of the chalk area of the Marlborough Downs are aptly named, in that they lie near to streams which flow only in the wetter periods of the year. Dry valleys may be produced in other ways than by river capture, and the reader is referred to pp. 132 and 135.

An escarpment may be modified very considerably by such stream developments, and there is obvious evidence of these changes to be found on maps of such land forms, and also by direct observation in the field. Fig. 77 shows a typical crest line of part of one of the chalk

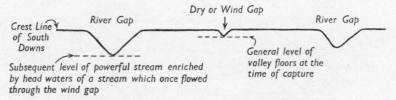

FIG. 77.—Gaps in an Escarpment, Generalized Profile of South Downs viewed from the North

escarpments in the South Downs of England. Two types of gap are shown, and each is of value in that it assists and controls communication by road and railway. Even the dry gaps, sometimes known as 'wind gaps', despite their relative height in comparison with the deeper-cut water or river gaps, ease the gradients which must be encountered in laying a route across an escarpment. Once again the wealden area is a very good one for the examination of these capture features.

Examples of River Capture

On the maps of an elementary atlas the student may discern the considerable river capture which has produced the system of the Yorkshire Ouse. Here there has been a succession of river captures by a powerful stream of the past which entered the sea by a mouth which corresponded to the present-day Humber. This stream captured first one and then another river on the north side of its course, which was roughly eastwards to the sea. Remnants of the captured rivers may be seen in the relatively short coast streams of the Esk and Rye, although today the latter flows into the River Derwent, which goes southwards out of the Vale of Pickering. Reversed drainage may be identified on the map in the district of the Hambledon and Cleveland Hills (Fig. 78). The pattern of the River Derwent is not explained on these lines, since its history is complicated by the effects of glacial diversion, which is discussed in Chapter VI, p. 109.

Furthermore, there is, without doubt, widespread evidence that there were numerous rivers flowing from north-west to south-east across the area of Wales, the Midlands, and South-east England, which by various river captures, assisted by local warping of the surface, have developed into the present-day patterns. It would seem that the headstreams of these rivers (which presumably fed, in the past, a river that is represented by the modern River Thames) were diverted in different areas to form the Lower Severn–Warwick Avon

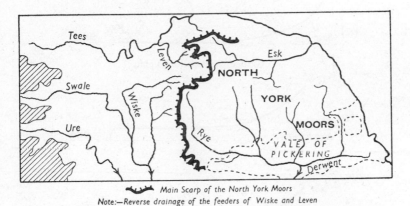

⌣⌣⌣ Main Scarp of the North York Moors
Note:—Reverse drainage of the feeders of Wiske and Leven

FIG. 78.—River Capture in Northern Part of the Basin of the Yorkshire Ouse
(*Based on the Ordnance Survey Map with the sanction of the Controller of H.M. Stationery Office*)

drainage, the Trent drainage, and the Wash drainage. In order that the reader may envisage this process more clearly and that he may accept this suggested history more readily, he should remember that the surface on which such a scries of streams was first developed lay at a much higher level than the present surface level of this part of the British Isles. There is evidence to show that there are clearly defined surfaces of planation throughout this wide area which give testimony of earlier river plains that have now very largely disappeared as a result of the general reduction of much of this surface by river erosion.

Rejuvenation

Two other features of some valleys are the entrenched meander course and the appearance of terraces on either side of the valley at various levels. These represent the effect of rejuvenation on the stream concerned. The details of the cycle of erosion have been given earlier. All streams are attempting to produce a graded course, and ultimately this would be achieved on one condition, namely, that no appreciable change in the relative levels of land and sea occurs. It is exceedingly doubtful whether a cycle of erosion is carried to its final stages because of the disturbances of level which are likely. Slight oscillations of the mean levels of both land and sea occur more frequently than those which cause large-scale marine transgressions, and it is possible to consider a localized change of level which may affect merely a section and not the whole of a river basin. The effect

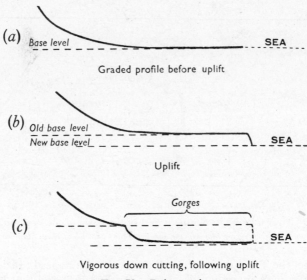

Fig. 79.—Rejuvenation
The stream may have two sections with their own base levels.

of uplift on a stream is to destroy its grade, and this often encourages the stream to resume the down-cutting which was typical of its youth. With the onset of fresh down-cutting the stream is said to be rejuvenated. The effect of such rejuvenation on the profile of the river is shown in Fig. 79. Some slight rejuvenation may occur in the upper parts of a river's course with an increase of rainfall.

Entrenched Valleys

The lowest section of a river course is usually in the form of a wide and low-lying plain, across which the main stream meanders on a grand scale. There are, however, rivers whose lower tracts are in quite elevated country with steep-sided valleys that resemble the immature upper tract. For instance, the River Wye, in southern Herefordshire and Monmouthshire, enters such a gorge-like valley after passing through Ross-on-Wye, and it does not leave it until reaching Chepstow at its entry into the estuary of the River Severn. The Wear and Tyne Rivers of northern England have similar, if less spectacular, gorges at their seaward ends. Other rivers have these narrow valleys in their middle courses, such as the River Rhine between Bingen and Bonn.

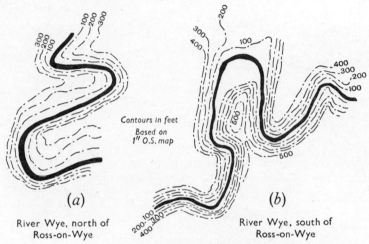

Contours in feet
Based on
1″ O.S. map

(a)

(b)

River Wye, north of
Ross-on-Wye

River Wye, south of
Ross-on-Wye

FIG. 80.—Entrenched Meanders
(Based on the Ordnance Survey Map with the sanction of the Controller of H.M. Stationery Office)

The lower Wye has an entrenched or incised valley which is as tortuous as was the course of the river over its flood plain before local uplift disturbed its grade and caused it to be rejuvenated. The valley sides descend to the river precipitously in some sections, as for instance at the Yat Rock, near Symonds Yat, giving a difference of level of almost 300 feet between the floor and the shoulders of the valley. Above Ross-on-Wye towards Hereford the river shows similar features, but here the valley never appears to be gorge-like, as it does farther to the south. Here a steep slope on one side of the valley is paired with a much more gentle slope on the opposite side showing that side cutting has played its part, but the cross-section of the valley is not typical of the lower tract of a mature river. The contrast between the Wye valley above and below Ross is shown by the contoured sketch-maps in Fig. 80. In some parts of the Wye valley in Monmouthshire there are crescent-shaped side valleys which have been abandoned by the meandering river after rejuvenation in much the same way that oxbow lakes are left by meander cut-offs in the flood plain. Fig. 81 shows an example of such an

Paul Popper

FIG. 81.—An Abandoned Meander Course, River Ebro, Spain
The central area, surrounded by the dry meander bed, shows radial drainage.

abandoned loop on the side of the Ebro valley in Spain. It is possible for such short-circuits to be caused by glacial diversion (see p. 110). Entrenched valleys may be found in areas where glaciation has left in its path a thick mantle of drift that has elevated the land surface, and thus caused the rejuvenation of the streams so that deep-cut valleys have been excavated in post-glacial times (Fig. 103).

V. C. Browne

FIG. 82.—Part of the Canterbury Plains, South Island, New Zealand

The rejuvenated valley of the River Waipara with its entrenched valley, braided course and estuary. Note also the bars along the shore enclosing lagoons and the shore flats leading inland to low cliffs.

The strategic positions afforded by the steep sides of an entrenched valley have been utilized by man. The older part of Durham City is enclosed within a great entrenched meander loop of the River Wear, so that its site was impregnable in the past apart from the narrow isthmus of land on the northern side of the town. On this side of the old settlement a castle and wall have guarded the northern approach (Fig. 83). The core of Berne, the capital city of Switzerland, is likewise enclosed within a bend of the River Aar. It is ·clear that in some instances the local uplift

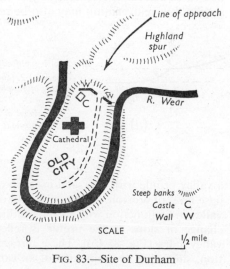

FIG. 83.—Site of Durham

which led to rejuvenation was so gradual that the river, with its renewed power for cutting down, was able to retain its grade whilst carving a gorge. Such was the case with the middle Rhine between Bingen and Bonn. There is a distinction between the two processes of entrenchment, as illustrated by the Wye and the middle Rhine, but the resultant valleys are essentially the same. Drainage which is successful in maintaining grade during a period of uplift is called antecedent. It will be seen that antecedent drainage is a form of superimposed drainage consequent upon local uplift.

Terraces

Terraces are marked by distinct breaks in the slopes of valley sides, and they are often paired, in that they lie at the same level on either side of the valley above the stream bed. On the other hand, it is possible for such a correspondence of level between opposite terraces to be absent. The distinction is shown in Fig. 84. The longi-

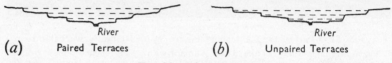

(a)   Paired Terraces   (b)   Unpaired Terraces

FIG. 84.—River Terraces

tudinal profile of some streams may exhibit a considerable divergence from the perfectly graded state. This lack of grading may be due to structural differences, as has been explained in a previous section, but rejuvenation may also be responsible. Fig. 85 (a) shows

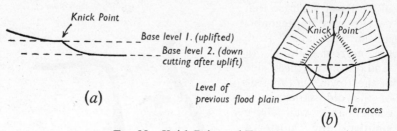

FIG. 85.—Knick Point and Terraces

a valley profile with two sections each of which has an independent base level of erosion. An examination of the cross-section of such a valley in conjunction with its longitudinal profile will show that this change of slope in the profile is reflected in the terraces on the valley sides (Fig. 85 (b)). These terraces are products of rejuvenation, since they are remnants of an earlier flood plain of the river.

Uplift has resulted in down-cutting, and the inner section of the previous flood plain has been removed. It is possible to trace terraces on each side of a valley until they converge at some point on the stream course which is known as a knick point (see Fig. 85). There may be several terraces with their associated knick points in one stretch of valley, each pair representing a remnant of a distinct flood plain of the past. These indicate that there has been more than one uplift with its consequent rejuvenation. These successive uplifts were in all probability separated by considerable intervals of time, when down-cutting attempted, not always with success, to restore the graded profile which had been disturbed. The geomorphologist regards each of these intervals as a distinct cycle of erosion, and each cycle is thus marked by a particular terrace line. The persist-

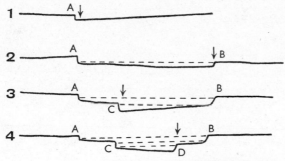

FIG. 86.—Development of Terraces by a Meandering River

ence of a terrace for great distances along a valley side is unusual, because of the breaching of the terrace by tributary streams, and in certain cases the swing of meanders in the development of a plain may destroy the terraces.

Terraces may be produced by other means than rejuvenation. It has been noted that terraces on opposite sides of a valley are not always coincident in level. The reason for this may be that the terraces have been produced by a meandering river as it swings across the flood plain and migrates downstream. It is essential to accept the possibility of some down-cutting by the river, even when meandering is well developed, in order to explain the formation of unpaired terraces. The composite diagram (Fig. 86) attempts to show the development of unpaired terraces in this way. The arrow in each stage of the diagram marks the position of the river as it moves with the progression of its meanders to different parts of the floor of the plain. The terraces have been formed with down-cutting in the order

A, B, C, and D. A series of terraces is a common feature in post-glacial valleys, where the work of the stream is concentrated in largely unconsolidated rock. Terraces are of considerable import-ance, because they are often the sites of settlements (their gravels and sands are frequently water-bearing), and they are valuable as lines of communication above the flood-plain level.

CHAPTER VI

GLACIATION AND GLACIAL LAND FORMS

Glaciers

During the recent geological past, ice from the north spread over much of Canada and the northern states of the United States of America, whilst from Scandinavia it moved south to cover the north of Germany, and from the Scottish Highlands and elsewhere it was dispersed over much of Northern and Central England. In this Ice Age the northern world was not continually covered with ice. Four times at least the ice advanced southwards, and four times did it retreat polewards. One obvious condition which promotes an advance of ice is that the prevailing temperatures of the air must be low, and it would appear from geological evidence that over this period of half a million years the temperature fluctuated so that at times the climate was relatively mild. Today we live in a northern world which witnessed the start of the last retreat of ice about 20,000 to 25,000 years ago. This is a small fraction of the 500,000 years which are allocated to the whole Ice Age, and it is possible that our mild climate of today is a mere interlude, and that in time it will deteriorate and there will be a fresh advance of ice from the north. In the northern world there is therefore abundant opportunity for observing the effects of an ice cover which has so recently been withdrawn.

Glaciers of today are of two main types, those which are continental and cover the whole surface of the land to its sea margins, as in Greenland and Antarctica, and those which follow valley routes in mountain areas and usually terminate long before they reach the sea, as in the Alpine regions of France and Switzerland and in parts of very high mountain zones in most areas of the world. The former group are essentially arctic in their climatic conditions, since low temperatures persist throughout the year and the ice does not melt until the ocean is reached. The latter or alpine type is usually found in the temperate latitudes, so that descent of the ice brings the glacier to levels within the mountains where melting is induced. However, such glaciers are found in the Himalaya and Andes mountains.

Formation of Glaciers

As a glacier moves it must be fed with ice at its head in order that it may persist, and it must be realized that low temperatures alone are insufficient; they must be accompanied by adequate precipitation of snow to maintain a constant supply of ice. The speeds at which glaciers move are very varied. Some may move more than fifty feet in a day, but the majority travel less than five feet in this time. It is clear that a glacier, while losing ice at its lower end by melting or by calving into the sea where large fragments float away as icebergs, must eventually disappear unless it gains at its head the equivalent amount of ice which is lost at its lower end or snout.

Glacier ice is produced from snow which falls during the year on the high parts of mountains or the inner regions of continental areas. Some of this accumulated snow may melt in summer, or it may evaporate and be lost as an effective agent for the production of ice. Even in sub-Arctic regions, such as Spitzbergen, there is a distinct loss of snow by summer melt, and the surface layers of glaciers turn to melt water. This ablation of snow is a factor which must be more than balanced by accumulation of sufficient snow to give an adequate supply of permanent ice to maintain the glacier. The alternation of accumulation and melting is shown on mountain slopes by the variation of the snow level or line which descends in winter and rises with the warmer conditions that prevail in the summer months.

An examination of large-scale maps of Alpine areas which possess glaciers today will show the frequent use of the word 'firn' (Fig. 87). This word is applied to the snowfields at the very head of a glacier and has entered into geographical literature. It is a German word, and there is a French equivalent in the term 'névé'. The German word means 'of last year', and the two terms, firn and névé, are used to define the collecting grounds from which glacier ice is derived, but the word firn is also used technically to describe the intermediate or transitional state between snow and compact glacier ice.

The firn zone is the surface of several layers which extend downwards to the solid compact ice that forms the main glacier. This upper part of a glacier may be regarded as stratified, and the layers are shown in Fig. 88, which represents, not to scale, a section from the surface of a snowfield down to the compact ice region. The two upper zones of snow and firn respectively are each subdivided by thin crusts of ice, and these represent annual banding which could be used, to a much more limited extent, in the way that tree rings are employed to date the age of a tree. The snow is much interspersed

with air, and each ice crust marks a layer of snow which has melted
in summer after the winter accumulation and has then frozen to ice
with the onset of the following winter. Progressive layers have accumulated with successive years, and in the deeper layers the individual snow crystals have lost their feathery tips, and have apparently
grouped themselves to form particles of firn which in turn become

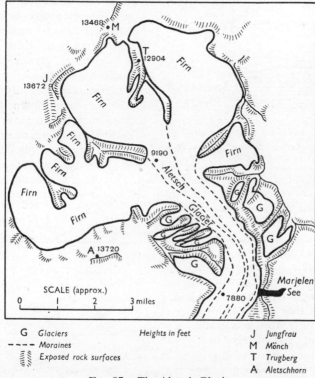

FIG. 87.—The Aletsch Glacier
(*Reference should be made to the frontispiece*)

coarser with increase of depth. There is a considerable amount of
enclosed air in the firn, but the proportion of solid content increases
with depth. The weight of the superior layers plays its part in com-
pacting the lower layers, and eventually a solid ice layer is formed at
varying depths from the surface. In one investigation[1] it was found
that with increasing depth the specific gravity of the layers increased.
Firn changed to solid ice when the specific gravity had risen from 0·6

[1] Made in 1938 by an Oxford University expedition on the Jungfraujoch in
Switzerland.

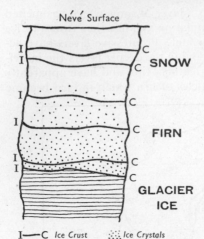

FIG. 88.—Vertical Section of Firn Zone
(not to scale)

I—C Ice Crust   :::. Ice Crystals

at the surface to 0·82–0·84 at depth. Even in the glacier ice there is still some enclosed air, but the crystals have massed together so as to exclude any possibility of intercommunication by means of fine air passages. Thus, whilst firn may be quite porous to melt water, which forms at the time of the summer heat, glacier ice is literally impervious.

Examinations of firn and glacier ice have shown that in the firn zone lateral movements of individual masses of ice crystals may occur within a given layer, but there is no comprehensive movement of the whole firn. On the other hand, glacier ice moves *en masse*, or, at least, in distinct layers. There appear to be thrust planes within the ice which permit differential movement of the several layers (Fig. 89). It is still impossible to render a full account of the movement of ice in a glacier, but it would appear that older theories concerning this movement are not completely overthrown by more recent investigations, which have confirmed to a certain extent the possibility of movement of crystals under stress in a liquid which is saline and derived from melt water. Sliding or shearing of adjacent layers is also very probable.

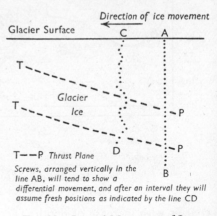

T——P Thrust Plane

Screws, arranged vertically in the line AB, will tend to show a differential movement, and after an interval they will assume fresh positions as indicated by the line CD

FIG. 89.—Lateral Movement of Ice
(*After May*)

Surface Features of Glaciers

In Alpine areas the firn lies in masses which are, in many instances, quite isolated; these frequently lie between exposed rocks which are either in the form of sharp edges or pyramidal peaks. However, many firn areas are found to coalesce at lower levels and give rise to a valley glacier (Figs. 87 and 90) which extends down-slope to its snout that is usually convex in front. The snout repre-

V. C. Browne

FIG. 90.—Franz Josef Glacier, South Island, New Zealand

The head of the glacier is seen with the firn zone in the background. Note the crevasses and the lateral moraine on the right of the glacier.

sents the level at which the rate of melting of the ice just equals the forward movement of the glacier and the melted ice disperses from it as a stream of water. The surface of the glacier is rarely smooth; it is usually fretted by deep cracks or crevasses which are largely created by strains that develop in the moving ice. The convex snout and the crevasses are largely due to the fact that the ice moves or slides downhill more rapidly in the centre than at the sides of the glacier, where frictional effects cause a retardation. The snout protrudes in the middle to accord with this effect. Crevasses develop at right angles to the direction of the strains which arise from the more rapid progression of the medial part of the glacier (Figs. 90 and 91).

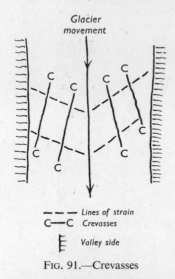

FIG. 91.—Crevasses

A longitudinal profile of a mountain glacier indicates that a very large crevasse, known as a bergschrund,[1] is formed near the head of the glacier in the firn zone; this particular feature is more obvious in the summer season (Fig. 92). The profile section also shows the crevasses and the depth of the glacier ice below the firn zone from which the former is fed. Glaciers in high latitudes, such as in Greenland and Antarctica, possess crevasses, but there is much greater continuity of the ice surface, since in these areas the whole land surface is literally buried by the deep ice cap and there is little friction at the upper surface of the ice. However, marginal areas of these continental ice covers are liable to stresses which result in the

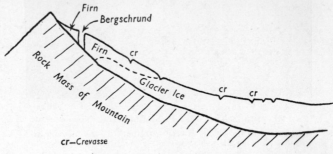

FIG. 92.—The Head of a Glacier
(*After May*)

production of a large number of crevasses, and, in addition, there are longitudinal crevasses lying approximately parallel to the direction of the ice flow. These arctic glaciers have no snouts, but where they reach the sea margins they form great cliffs of ice from which

Also known as a 'rimaye'.

icebergs detach themselves. In some instances the ice extends out to sea, floating on the water like an apron.

The Work of Ice

In areas from which a previous ice cover has retreated it is possible to estimate the work that has been done by glaciers. Glaciation has left in highland regions, such as those of North Wales, Scotland,

Standard Oil Co. (N.J.), photo by Webb

FIG. 93.—Glacier near the Susten Pass, Switzerland

Note the lateral moraine to the left and the terminal moraine, damming a lake to the right of the foreground. In the background is a corrie with a hanging valley to the right.

Scandinavia, and elsewhere, sharp-angled ridges or arêtes (Fig. 38), and pyramidal peaks such as the Matterhorn in the Alps of Switzerland. These two features may be regarded as developments from the precipitous-sided concave hollows known as corries or cirques (Fig. 38), which are almost ubiquitous in these glaciated uplands. In addition, there are to be found the U-shaped valleys (Fig. 103), with relatively level floors but almost precipitous sides, into which tributary valleys enter at high levels and thus form hanging valleys (Fig. 93). Very often extensive sections of these

U-shaped valleys exhibit a longitudinal profile which is scalloped (see Fig. 100), with present-day rivers following courses in narrow ravines and plunging over waterfalls at the points marked X in the diagram.

In the earliest days of glacial study it was considered that great ice masses acted like a chisel, or more probably as a gouge, scooping out or scalloping the landscape. It is clear, from the study of existing glaciers, that on the margins of the ice mass there is much weathering of the nearby rock surfaces. The action which produces this weathering is often referred to as sapping, and is consequent on the melting and freezing of ice by day and night respectively on the margins of cavities in the ice, such as a bergschrund or crevasse. Rock debris so produced, together with rock waste that falls from the slopes above, forms long trails of rock fragments that are known as moraines. Those which lie along the outer margins of the glacier are known as lateral moraines (Figs. 90 and 93), but where two or more

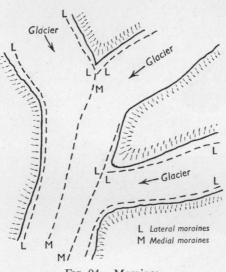

Fig. 94.—Moraines

glaciers unite, the confluent lateral moraines of each produce a series of inner moraines. These are called medial moraines from their positions (Fig. 94 and Frontispiece).

The base or sole of a glacier is a zone in which similar sapping takes place, because the under surface of the ice is readily melted owing to the great pressure which is exerted on these deep layers by the overlying ice mass. Frost action will break up the underlying rock surface, and large fragments of rock are enclosed within the sole of the glacier. As these fragments move with the ice, they erode very powerfully and give the action which is likened to gouging. Issuing from the snout of a glacier is a stream of water which is not fed by the melt of the snout ice alone, but is, in fact, a subglacial stream which deposits in front of the snout its burden of rock fragments. These rocks, together with additional material from the lateral and medial moraines which are carried to the snout end of the glacier and there deposited on melting of the ice, form a third type of moraine known as a terminal moraine. Recession of the glacier by prolonged melting, with warmer conditions of climate, leaves the land surface, which is thus exposed, littered with rock fragments, and these constitute a ground moraine. It is this moraine which lies over such extensive lowland areas adjacent to glaciated highlands, although much may be found on the floors of the mountain valleys.

Despite the erosive power of the enclosed fragments of rock in the sole of a glacier, many geographers have been sceptical of the potency of ice erosion, since ice can be deformed by the irregularities of the surfaces over and against which it passes. Some geographers have argued that ice has acted as a protective cover to the land surface, and that it was the areas not so covered which were actively eroded in the Ice Age by running water. The outstanding forms of glacial landscape will now be discussed more fully, and regard will be paid to the fact that some authorities believe in the effectiveness of ice erosion, whereas others consider that water erosion outside the ice-protected areas has been the more important agent.

Corries

In many of the Welsh mountains and in many parts of the Scottish Highlands the outstanding peaks are surrounded by quite deep hollows or corries which have arcuate margins enclosing areas often more than semicircular in plan (Fig. 95). These hollows have precipitous walls that may tower as much as a thousand feet above the floor of a depression, which is not infrequently occupied by a lake whose outlet is into a typical U-shaped valley (Fig. 96). The summit of Cader Idris in North Wales is thus surrounded by several corries. These depressions were probably initiated by the process of nivation. Nivation depends upon the collection of snow in a hollow, or at a valley head, and on its margins there is thawing and freezing in turn which has the effect of sapping at the adjacent rock surfaces, with the result that there is developed a very powerful form of mechanical erosion. The alternating thaw and freeze may exhibit a diurnal as well as a seasonal rhythm. Long-continued action of this kind results in a recession of the inner walls of the hollow. With greater accumulation of snow as in the firn areas of the Alpine peaks and with the formation of a marked bergschrund, the sapping would be much accelerated. Later, with the formation of glacier ice and its enclosed rock fragments, the hollow would be deepened.

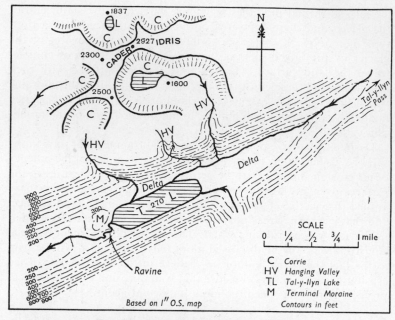

FIG. 95.—Cader Idris Region, North Wales

(Based on the Ordnance Survey Map with the sanction of the Controller of H.M. Stationery Office)

Arêtes and Cols

Knife-edge ridges may be developed when the walls of two back-to-back corries have met after sapping from both sides. Fig. 96 (*a*), shows a north to south section of the summit region of Cader Idris. Clearly it would involve very little more erosion to produce a knife-edge ridge or arête as shown in Fig. 96 (*b*). Striding Edge, on Helvellyn, in the Lake District, is an admirable example of an arête. Several corries whose walls have receded by erosion on various sides of a mountain summit in time create a pinnacle which is pyramidal in shape, with a series of arêtes rising to a culminating 'horn' as

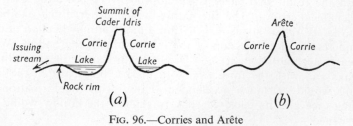

FIG. 96.—Corries and Arête

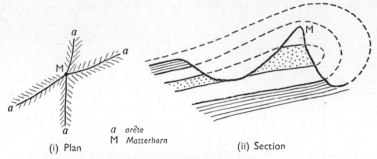

a   arête
M   *Matterhorn*

(i) Plan   (ii) Section

FIG. 97.—The Matterhorn

The Matterhorn has been derived from a recumbent fold which has been partially reduced by ice action, leaving a pyramidal peak.

shown *par excellence* by the Matterhorn and many other Swiss peaks (Fig. 97). The reduction at a later stage of an arête gives rise to a col, one of the most important features in high-mountain terrain since it often provides the means of intercommunication in such areas (Fig. 98). It must be noted that a col may be formed in other ways and does not necessarily depend upon glaciation.

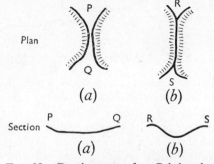

FIG. 98.—Development of a Col by the Intersection of the Walls of Two Adjacent Corries

(*a*) Just before intersection.
(*b*) After intersection.

U-shaped Valleys

The U-shaped valley is regarded by some authorities as wholly the product of ice erosion. Through a V-shaped river valley of pre-glacial origin a glacier has forged its way almost as a gouge, so that the profile of Fig. 99 was developed. The erosion was accomplished partly by sapping and partly through planing by the ice, armed with large rock fragments. The scalloped floor, already referred to, can be explained by differential erosion, and the steps in the longitudinal profile (at X, Fig. 100)

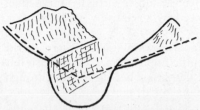

FIG. 99.—A Glaciated Valley. U-shaped Valley imprinted into V-shaped Valley

are interpreted as marking zones where the rocks were more resistant to erosion than above and below these points. In this respect the steps in the valley floor may be compared with the ungraded sections of the immature tract of a river's course. On the other hand there is a strong likelihood that at these points ice from confluent glaciers joined the main valley glacier from the sides, and produced a great increase in the erosive power of the glacier, resulting in these irregularities in the floor of the U-shaped valley.

From the point of view of ice protection the U-shaped valley was formed in the following way. The upper part of a pre-glacial river valley became occupied by ice at the onset of cold conditions, whilst down valley, below this ice cover, normal water erosion proceeded. On the retreat of the ice, the valley would show two distinct sections, an upper one which, owing to ice protection, would be relatively ungraded compared with the lower part that had been continuously

FIG. 100.—The Scalloped Floor of a Glaciated Valley

developed by running water. Upon the resumption of water erosion in the upper part of the valley, vigorous down-cutting would take place, and a deep V-shaped valley would be developed. Advance of the ice at a later stage would give protection to the upper section of the valley, although it is likely that some modification of this valley would result from the movement of the glacier ice, so that its floor would be widened to a certain extent at least. Repetition of this advance and retreat of the ice with alternating periods of water erosion and ice protection would result eventually in a deep U-shaped valley which had not depended solely upon ice erosion for its excavation. The scalloped floor of the U-shaped valley would be almost inevitable, since complete grading by successive periods of water erosion in front of the ice cover would be virtually impossible. Each halt of the ice would be marked by a relatively graded section of the valley (Fig. 101).

In many of the exposed glaciated valleys of the northern world there is evidence of valley spurs on the higher slopes. At lower levels these spurs have been reduced at the time when ice moved through the valley and widened it by planing off irregularities on its sides. Consequently, the basal sections of such spurs are truncated, and they fall in line with the markedly rectilinear walls of the glaciated

valley, which are unbroken, save where today tributary streams enter
through V-shaped notches that are being cut down with great rapid-

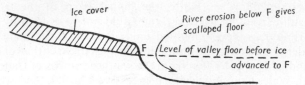

FIG. 101.—Development of a Scalloped Valley Floor

ity by very youthful streams (Fig. 102). The remnants of the spurs,
above the valley walls, are proof of previous water erosion, and
whilst the ability of ice to clear the sides of such a valley is admitted
by most geologists, many hold the view that its depth has been determined primarily by rivers
and not by ice.

Hanging Valleys

The hanging valley is one
of the other interesting features in glaciated regions.
Usually a tributary river
joins its main stream at the
same level, in other words,
the two valleys are accordant. It is only in the ungraded mountain tract of a
river valley that it is possible
to have a tributary plunging
into the main stream over
a waterfall. In mountain
valleys of glaciated regions it

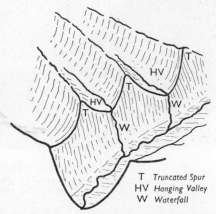

FIG. 102.—Truncated Spurs and Hanging
Valleys

is very common to find a tributary stream descending by falls and cas-
cades for several hundreds of feet down the valley sides before it enters
the main stream (Fig. 102). Neighbouring tributary valleys do not
necessarily show the same amount of discordance between their floors
and the floor of the main valley; in fact, some may be quite accordant.
This discordance has exercised the minds of geographers and there is a
variety of explanations. Those geographers who believe in the effectiveness of ice erosion attribute the discordance to the unequal excavation by a major glacier and its more shallow, and therefore less
effective, tributary glaciers. Those who believe in ice protection
maintain that most of the deepening must be fluvial and that it has

occurred in inter-glacial periods. To explain fully the hanging valleys, it is suggested that in a period of ice retreat the main valley was unoccupied by ice, whilst the higher parts of the tributary valleys still carried an ice cover and could not be eroded by water. These tributary valleys remained ungraded while the main valley was undergoing active erosion. The various streams that enter the valley and lake of Tal-y-llyn from the southern slopes of Cader Idris in North Wales hang in varying degrees (see Fig. 95). Since the final

Judges Ltd.

FIG. 103.—Tal-y-llyn Pass, Merionethshire

Here is seen a U-shaped valley with the post-glacial development of a V-shaped valley in its floor. Tal-y-llyn Lake can be seen in the distance with its deltaic flats (see Fig. 95).

retreat of the ice, considerable river work has notched the upper part of the U-shaped valley of Tal-y-llyn with V-shaped ravines, but the side streams are far from being graded (Fig. 103).

Roches Moutonnées and Erratics

Within a glaciated region it is usual to find rock surfaces which are grooved and scratched. These striations have been made by angular rock fragments enclosed within the soles of glaciers, and they give some indication of the direction of ice movement, a matter of geological importance. Some of these rocks are known as roches moutonnées. In section a roche moutonnée shows two distinct aspects of surface: (*a*) a smooth surface with a slight slope, and (*b*) a relatively irregular surface with quite a steep slope (Fig. 104). The two surfaces represent different actions of moving ice, the former indicating the planing action, combined with scratching, whilst the latter slope is the result of the plucking and sapping action similar to that which may be observed in the bergschrund zone of a glacier. Fig. 104 shows how the ice has moved over these rocks and that the plucked surfaces are on the down-flow sides.

Erratics may also be found lying on the uplands, and for that matter in the lowland areas beyond the mountains over which ice has moved. For instance, boulders of granite, unmistakably derived from a granite area in the Criffel district of the Southern Uplands of Scotland can be found in fair abundance in the South and Central Pennine regions of England. These fragments have been transported by ice in a glacier that moved at some time from Scotland in a southerly or south-easterly direction. By tracing erratics

FIG. 104.—Roche Moutonnée

back to their source it is possible to ascertain the direction in which the ice flowed, and where erratics have been found to contain valuable ores of metals it has sometimes been possible to locate the ore body. An erratic is shown in Fig. 123.

Glacial Lakes

Throughout glaciated regions there are innumerable lakes of various types which owe their origin, largely, to glacial action of one sort or another. The base of a corrie is hollowed out by ice action, so that there is very often a rock rim at its foot which impounds a corrie lake or tarn (Fig. 96). Many of the very small tarns which may be said to litter the surfaces of glaciated mountains occupy very minor depressions that have been scooped out by ice.

The lake at Tal-y-llyn, near Cader Idris, has a somewhat different history. The valley in which this lake is situated is a deep U-shaped one, and immediately below the lake there is a ridge which rises several feet before plunging down some hundred feet or more in a distance of about a quarter of a mile (see Fig. 95). The stream which drains Tal-y-llyn lake makes its way through a ravine that it has cut on one side of this obstructing ridge. The ridge is a terminal moraine which lies athwart the main valley at this point where at some time a glacier, which occupied the main valley, had its snout. Such a lake is moraine-dammed, and there are many examples to be found throughout all glaciated regions. Tal-y-llyn lake was probably much larger at one time, for its upper end has been infilled with deltaic deposits.

Long narrow lakes, which are also of very considerable depth, may occur in conjunction with the scalloping of a glaciated valley. Such over-deepened basins are frequent in Northern Europe and in particular in the Swedish mountains, where almost every valley which descends from the Kiolen range to the Baltic Sea possesses one of these lakes, often referred to as a glen lake. The majority of the Cumbrian lakes are of this type.

Lakes dammed by Glaciers and their Outflows

Today, there are valley glaciers which arrest the flow of a side stream and produce lakes. The Marjelen See, on the east side of the Aletsch glacier in Switzerland, is of this type (see Fig. 87). Such present-day lakes have suggested that there must have been quite extensive lakes of this type at the time of the greatest extension of the ice sheets in the northern world, although the lakes are no longer in existence. Evidence of such lakes is available in the region of Glen Roy and its associated valleys, off Glen More and some distance north-east of Ben Nevis in Scotland. The lower ends of various pre-glacial valleys were choked with ice, and lakes were impounded in the upper parts of the valleys. The level of each lake rose until a height was reached which permitted its water to flow over an inter-vening ridge and obtain an outlet through a valley which was not affected by glacial damming. It will be remembered that the ice cover fluctuated considerably during the Ice Age, and such a system of lakes would be modified as the ice shrank with a change of climate; the level of the lakes fell, and when the ice halted in its retreat, a new outlet or series of outlets was established. In this area of Glen Roy there are no less than three significant levels on the valley sides which are occupied by beach deposits of the pre-existing lakes; they are known as the 'parallel roads' of Glen Roy. The levels of these de-posits coincide with the heights of well-defined cols or spillways through which the lake waters found an outlet at various times.

In North-east Yorkshire there is abundant evidence of such glacial damming. The probable pre-glacial relief is shown in Fig. 105 (a). Ice advanced from the north, and was arrested by the uplands of the wolds and moors, so that several lakes were formed by streams that had flowed eastwards to the sea. These lakes, like those in Glen Roy, spilled over at convenient levels, and the out-flowing streams cut channels which remain today, one of which is followed by the railway between Pickering and Whitby (Fig. 105 (*c*)). The small northern lakes probably had a common drainage through the moors into a larger lake which spilled southwards into the Vale of York (Fig. 105 (*b*)). Today all the lakes have disappeared, but the south-eastern part of the York moors is drained southwards by the Derwent, which flows through Forge Valley, yet another spill-way, and then across the drained bed of the largest lake, now represented by the Vale of Pickering, and finally into the Vale of York by

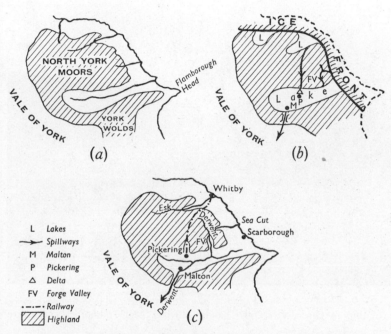

FIG. 105.—Glacial Lakes and Spillways in the North York Moors

way of the Malton gorge (Fig. 105 (*c*)). It may be noted that Pickering stands on a delta constructed by a stream which entered the large lake of the past (Fig. 105 (*b*)).

Other examples of such diversions may be found throughout Northern England and the Midlands (Fig. 106). The Ferryhill gap between Darlington and Durham has been cut in the Magnesian Limestone escarpment which once separated two glacial dammed lakes. The lower Wear has a gorge-like valley near Sunderland which the river cut when its earlier course into the Tyne near Gateshead was obstructed by ice. Ice from the Irish Sea deflected the Severn (which flowed in pre-glacial days northwards into the sea near the present Dee estuary) to the east, where the ridges of the Wrekin and Wenlock Edge caused a large lake to form in the Shrewsbury area; this spilled southwards and thus the Severn gorge was cut between Buildwas and Bridgnorth and consequently the lake was drained. Near Llangollen the River Dee has abandoned some earlier-formed meander loops as a result of glacial action. The loops were plugged

Geological Survey photograph reproduced with the permission of the Controller of H.M. Stationery Office

G. 106.—A Glacial Overflow Channel, Cliviger, near Todmorden, Yorkshire

by ice and glacial deposits, but the Dee was able to cut a new and straighter course, which in post-glacial times it has retained.

Lakes formed by Glacial Warping

At the time of the maximum ice cover it is estimated that the thickness of the ice was in some places 6,000 feet, although, as with the cover in Greenland today, it was probably lens-shaped, and therefore tapered considerably towards the margins (Fig. 107 (*a*)). Nevertheless, it is clear that a great weight was imposed upon certain areas which was of sufficient magnitude to create a down-warping of the crustal layers of sial. This down-warping may be compared with the depression of a geosynclinal area (as envisaged in the formation of mountains), but on a much smaller scale. There is evidence that Scandinavia and the Baltic basin, as well as the Canadian Shield, are rising as a result of the restoration of isostatic equilibrium following

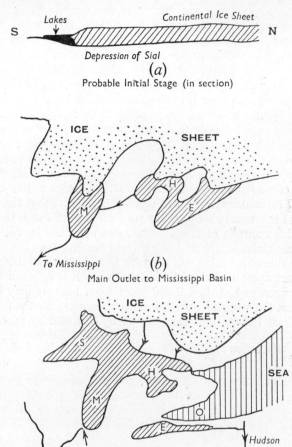

(a)
Probable Initial Stage (in section)

(b)
Main Outlet to Mississippi Basin

(c)
Main Outlet to a Sea in the East

FIG. 107.—Stages in the Development of the Great Lakes of North America

the unloading by the ice melt at the end of the recent glaciation. This elevation of the areas in question is still in progress, partly because the unloading is not complete (there are remnants of the ice sheets on the highlands of Scandinavia which are still shrinking) and partly because the uplift has not kept pace with the melt of the ice.

At the time of the main retreat of the ice at the end of the Ice Age, land surfaces were exposed, but these remained sufficiently down-warped to encourage the formation of fairly large lakes. Some of these marginal areas were flooded by the sea. It was under such circumstances that the Great Lakes of North America had their origin, although there is evidence that the hollows, created by down-warping, were also scoured, to some extent, by ice. The lake basins were undoubtedly formed on the margins of the ice sheet that covered northern North America. The earliest of these lakes were able to spill southwards towards the Mississippi basin, but as the ice receded northwards the lakes increased in size and found outlets in various directions. At different times the lakes drained towards the Mississippi, the Hudson valley and an arm of the sea, which occupied an area corresponding to the lower St. Lawrence valley of today (Fig. 107 (b) and (c)). Of great interest is the fact that the modern Great Lakes are just able to maintain their eastern outlet because of the existing controls of uplift consequent upon the ice melt, but if the present trend of isostatic adjustment persists, then uplift on the northern side of the lakes will permit Lake Michigan to drain southwards once again into the Mississippi basin. It is believed that this could happen in less than 2,000 years from now. It may be mentioned that this warping does not account entirely for the formation of the Great Lakes, since they are partly dammed by moraines.

Lowland Glaciation

At the time of the most intense glaciation in the Pleistocene period it is probable that many areas were covered completely by thick ice sheets equivalent to those in Greenland and Spitzbergen today. Clearly valley glaciers occupied many highlands of the northern world, and these extended to the lowlands nearby, where they spread out and joined with one another to form great piedmont glaciers. Such glaciers had little opportunity of gathering much surface moraine, since at the time of the maximum glaciation most of the land surface in the mountains was buried by ice. However, such continental ice sheets did create on their under surfaces what are known as ground moraines, which have been left as a cloak over the preglacial land surfaces. At the same time there was a continuous outwash from the margins of these ice sheets where melting was constantly taking place.

There remain in many areas of the European and North American lowlands great terminal moraines which are the melt deposits that accumulated in front of the ice sheets when they remained stationary for a fairly long period of time. These moraines indicate that there were several prolonged halts during the final retreat of the ice. The North German plain is crossed by many parallel, arcuate lines of moraines, and similar groups occur in other lowland regions such as those which lie both east and west of the Pennine Uplands in England over which ice was dispersed. Most of these deposits are clays with enclosed rock fragments which vary in size and quantity from locality to locality. The names of boulder clay, glacial drift, and till are applied to these formations (Fig. 136) which in parts of Eastern England are of such a depth that if they were stripped from the underlying rocks they would cause the present coastline of East Anglia to recede by many miles. A study of these deposits is of importance to the geographer because, usually, these clays are of great fertility, in which case they influence farming activities to a high degree.

Ground Moraines

In front of most of these terminal moraines which lie in the lowlands there are very extensive outwash plains of clays and gravels, whilst behind each moraine it is quite common to find a group of

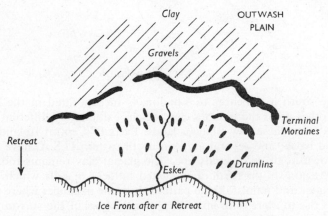

FIG. 108.—Ground Moraine and other Deposits

hummocky low hills, known as drumlins, and winding low ridges, known as eskers and kames (Fig. 108). The outwash plains are made of sands and gravels, often interbedded with boulder clay. The gravels represent, in general, material that has been carried forward either by melt water dispersing from the ice or by rivers which have affected these regions in post-glacial times. The term 'fluvioglacial' may be applied to these deposits. The boulder clays are ground moraine deposits, and may sometimes lie above gravels, in which case the latter have been overridden by a later advance of the ice. Unfortunately, it is not always easy to distinguish the clays and gravels, since both may be modified locally to such an extent that the boulder clay is largely a deposit of boulders and stones without clay and the gravels may be intermixed with clay.

In Eastern England large level drifts of boulder clay form dissected plateaus of no great elevation, and it is considered that these deposits were derived from ice which remained stationary after its final advance. The clays contain boulders from distant areas, but also much local rock, such as chalk in the areas of North Essex, whilst London Clay (a rock of pre-glacial formation),[1] lying farther

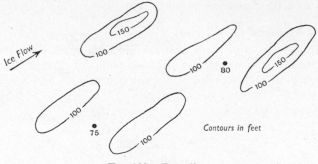

FIG. 109.—Drumlins

to the south and nearer to London, is incorporated in the drift.

In other areas the boulder clay regions exhibit very different land forms from these low plateau lands. Parts of Central Ireland, the Tweed basin, and several areas in north-eastern U.S.A. have what is known as drumlin landscape. The glacial clay remains today in these regions in the form of elongated hillocks or hills which form a peculiar and unmistakable pattern. In some instances where these ridges rise to more than fifty feet above the level of the surrounding land, it is possible for the pattern to be discerned on the large-scale maps of the district by almost elliptical contours (Fig. 109). These ridges are usually found behind terminal moraines and it seems that they have been developed in the ice mass and left upon its retreat. The map shows that they are arranged longitudinally with their longer axes in parallel lines, and this is in sympathy with the direction of ice-flow movements which affected the region. The

[1] Of Eocene age.

drumlins are not in alignment when viewed transversely, which suggests that they were affected individually by the moving ice. Possibly they are great masses of ground moraine which were engulfed by the lower layers of the glacier ice and yet remained practically stationary despite the movement of the ice in general.

Eskers and Kames

Here a brief reference must be made to eskers. These are sinuous gravel ridges which run across country often quite regardless of valley lines or hills. They are clearly of river origin, and are thought to represent the deposits of subglacial streams which threaded their ways in tunnels under, or in, the bottom of the ice sheets. Many glaciers of the Alpine region show outlet tunnels of this nature with an issuing stream. It is also likely that eskers may be deltaic fans that were formed at the mouths of such tunnels where they

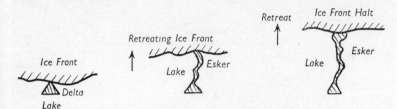

FIG. 110.—The Formation of an Esker

had an outfall into a lake which was marginal to the ice front. On the recession of the ice sheet the delta extended backwards in the form of a narrow band of deposits, since it has little opportunity to spread out into a fan. A halt in the recession of the ice caused an increase in the deltaic deposits which is shown by a swelling in the otherwise narrow ridge. Several halts of the ice are marked by a series of such swellings, and the esker is then said to be beaded (Fig. 110). Kames may appear in the same areas as eskers. They may be isolated masses of deposits, or they may be clustered in the form of a winding ridge. It is probable that kames are also of deltaic origin.

The eskers and terminal moraines may be of value to man when they form slightly elevated ridges in a lowland region, as they are usually high enough to escape both damage and interference from river floods. Consequently routes may follow the crest of an esker or a moraine. In the Central Plateau of Finland eskers which wind across lake basins, with sufficient elevation above the water-level, are invaluable in that region as lines of communication. One of the finest examples of this use of an esker is afforded by Punkaharju (Fig. 111). The town of York has grown up by the River Ouse on either side of a shallow gap that has been cut by the river in an east–west morainic ridge which crosses the Vale of York and gives a relatively dry causeway in time of floods. Some of the North German rivers have been diverted by the arcuate terminal moraines which cross the northern plain.

Crag and Tail

A rather special feature of glacial landscape is known as 'crag-and tail'. Reference has already been made to volcanic crags (p.

E.N.A.

FIG. 111.—Punkaharju, Finland

Esker ridges, above a lake surface in Finland, covered with coniferous forest.

30), and in the case of some such crags in the Lothian area of Scotland, particularly that on which Edinburgh Castle stands, there is a 'tail' to the crag. Fig. 112 (*a*) and (*b*) will help to explain the origin of this combined feature. The crag has been an obstacle in the path of ice which has moved over the area, causing the ice to pass over and round the crag. The tail is built of local sedimentary rock which was largely protected from the ice by the crag, but not sufficiently to prevent some glacial planing and some deposition of boulder clay, mainly on its upper surface. The sides of the crag which were exposed laterally to the ice flow are excessively steep, and give an al-

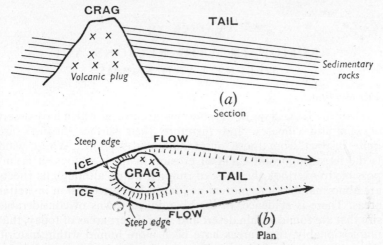

FIG. 112.—Crag and Tail

most impregnable position to Edinburgh Castle. Today the 'tail' carries the street of Canongate, which leads from the Castle to Holyrood House, lying at its end.

CHAPTER VII

EROSION IN ARID REGIONS AND DESERT LAND FORMS

Introduction

There are large areas of the continental surfaces which have desert or semi-arid climates, and therefore their surface features are quite distinct from those of more humid regions. Whilst wind is the most important agent of erosion in these regions, it is impossible to overlook the work of running water, although its effects are abnormal and intermittent as judged by river erosion in wetter areas. There is evidence, in the abandoned towns of considerable size that are found within desert and semi-desert areas of today, that unquestionably, these areas have been more humid within historic times. Nevertheless, although water erosion has been responsible for some of the landscape features of deserts, we must consider wind erosion of very great importance. The most effective and extensive action is that of deflation, whereby the land surface is lowered by wind action. Deflation is the blowing away by wind of weathered rock fragments. Thus both transport and accumulation of rock waste within and beyond arid regions may be almost exclusively æolian. Furthermore, evaporation and the lack of vegetation are other factors which affect the land forms of these regions. Most arid regions have no external drainage to the oceans. De Martonne, the great French geographer, has shown that nearly a third of the land surfaces of the world are characterized by internal drainage, but it must be realized that this type of drainage is by no means restricted to areas with a desert climate.

Deserts

Since the major areas of aridity are deserts, it is desirable to classify their various surface features. The desert surface may be of bare rock as in the hammada of the Sahara, where deflation is predominant. On the other hand, the erg or sandy desert represents a surface of deposition (Fig. 208). There is yet a third type of surface consisting of gravels and stones, and this is known as the reg in the Sahara. Some desert areas, such as those in western parts of the United States and in Tibet and Mongolia, are deserts largely because of tectonic features which have resulted in the formation of great depressions within the mountain zones (Fig. 234) that are characterized by low rainfall (because of the sheltering effects of the surrounding mountains) and by inland drainage. Although abrupt relief features of tectonic origin are found in the tropical deserts, more often these latter deserts coincide with structural units of plateau character, as in Western Australia and the Sahara.

Wind Erosion

The deflation process is, in a sense, the end of a series of processes which are largely mechanical, although some chemical weathering is of assistance. After the initial weathering of rock surfaces by exposure to both intense heating and chilling, the wind, armed with rock fragments, is able to abrade. In their turn the fragments produced by abrasion are worn down by their attack which is like a

ZEUGEN   YARDANGS

FIG. 113.—Zeugen and Yardangs

sand blast on the solid rock against which they are blown, and so great accumulations of rock debris are produced. As with a river so with wind, the load it carries must be adjusted to its speed, and it is not always possible for fragments of all sizes to be transported. The pebbles and gravels are not so readily removed as the finer sand which forms dunes within the desert areas, or is blown beyond the regions into more humid areas. There is clear evidence that occasional rain-storms are of value in providing alluvium of various grades for transportation by the wind. The almost constant erosion by wind is more emphasized in these regions than it is in humid areas. Thus the tabular zeugen and the undercut, parallel ridges, known as yardangs, may be developed (Fig. 113). Wind can lift fragments of sand into the air, and its most effective abrasion occurs at some distance above ground-level, as there is considerable frictional drag in the lowest layers of the air.

Dunes

The deposition of sand is usually in the form of dunes. It is possible to study dune formation on the sea coasts of humid lands, but there the dunes are somewhat peculiar in that they are readily fixed by vegetation. In addition they are usually on a small scale, since they have only a limited supply of sand. However, coastal dunes do show some of the features which are typical of desert dunes and they may be studied with profit wherever they occur in Britain and elsewhere. Desert dunes may be divided into two groups. Firstly, there are the barchans, which are nearly always transverse to the prevailing wind direction. The front of each dune is convex, and its leeward side is concave, with two crescentic ridges or horns curving away and sloping down from the crest at C (Fig. 114 (*a*)). In profile these dunes show a relatively gentle windward slope and a leeward slope which is quite steep and often concave as a result of wind eddies (Fig. 114 (*b*)). Secondly, there are dunes which form long lines that are parallel to the wind direction. These are secondary in

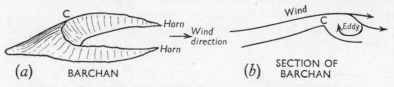

FIG. 114.—Form and Development of a Barchan

formation, in that they seem to develop from barchans or transverse dunes.

The barchans in true desert areas will move or migrate, so that the surface of a sandy desert may be compared to that of the sea with its progressing waves, except that the propagation of the dunes is much slower. But even in the very arid deserts there is rarely a complete absence of either plant life or of a water table. Thus certain barchans may become fixed at their horns by plants or by water which is near the surface. The water makes it impossible for the sand to be lifted (sand must be dry before wind can move it) and the plants act as a wind break. It is only possible for such fixation of dunes to occur at points where the sand has no great depth. Elsewhere the sand is so deep that both plants and water are buried by an overburden of sand whose upper layers can still be moved. The usual result of the fixation of a barchan is that a blow-out occurs when the centre of the dune is breached and its sand is strewn out in the form of longitudinal dunes in between which there is usually exposed a surface of bare rock (Fig. 115).

The Qattara depression in North Africa shows on a large scale the effects of deflation. It is over a hundred miles in width, and in parts it is over 1,000 feet below the level of the surrounding limestone plateau and has been excavated largely by wind, although it is probable that chemical solution has occurred as well. The prevailing winds are northerly (Etesian winds from the eastern

Mediterranean Sea), and south of the depression are longitudinal dunes that have been built of sand derived from the deflation and which extend for scores of miles. Neighbouring depressions have

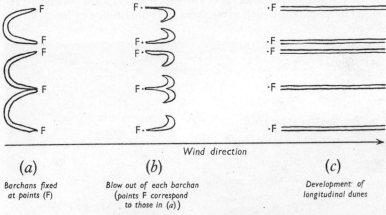

Wind direction

(a)   (b)   (c)

Barchans fixed   Blow out of each barchan   Development of
at points (F)   (points F correspond   longitudinal dunes
                        to those in (a))

FIG. 115.—Blow-outs in Barchans

been lowered in this manner to the level of the water table of the area, so that a series of oasis populations has been maintained by a water supply right up to modern times.

Water Erosion in Arid Areas

There are other characteristic land forms in arid regions which must be reviewed in brief. These show the effects of both regular and intermittent fluvial erosion of the past and the present day, respectively. In discussing these features it is necessary to make some distinction between those in plateau areas and those in the tectonic or intermontane basins.

Plateau surfaces in arid regions are scored with numerous ravine-like valleys, some of which show the interlocking spurs that are so typical of youthful river valleys in other regions, and the surface is

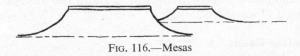

FIG. 116.—Mesas

thus dissected into the land form known as 'bad lands'. When a particularly resistant and horizontal layer of rock has capped a plateau its remnants may be tabular and steep-sided, in the form of mesas (Figs. 116 and 117). The sides of the intervening ravines are quite ungraded, although their floors may be remarkably wide and fairly well graded. Water transport has been limited largely because there is so much fragmentary materal to dispose of and the short-lived streams (since evaporation is so great) soon become over-loaded to an extent which gives them the consistency of liquid mud.

Black Star

FIG. 117.—'Bad Land' Surface near the Dead Sea, Palestine

Here are seen wadis with their ungraded sides, but showing evidence of water erosion. The mesa-like plateau surfaces show the effect of erosion on horizontal strata of varying resistance.

In the areas with extensive plateau surfaces the action of rain is to produce movement of material which is akin to sheet erosion of soil. The main result of water erosion is to develop great alluvial spreads over the floors of the valleys, together with lines of deposits which may be compared in their outward form with the morainic ridges left in front of ice sheets.

Intermontane Basins

In the arid regions of intermontane basins there are additional modifications to this picture. The steep fault scarps on the edges of these basins have remarkable screes and cones, built of the larger rock fragments which have been deposited by streams as their speed has been checked by the sudden change in gradient at the scarp bottoms. Towards the centres of these basins, beyond the marginal cones, other deposits are laid in a sequence which is related to the size of the fragments, the smallest being carried farthest away from the scarps. Thus a fringing zone of piedmont gravels is developed which surrounds an inner region of each basin, where fine material, chiefly sands, and sometimes saline deposits are found (Figs. 118 and 119). Drainage of the past or the present has produced a pattern which is the inverse of radial drainage from a dome. Streams or dry courses (wadis and arroyos) converge towards the central area which is occupied by a salt lake or a salt plain (or playa), both, of

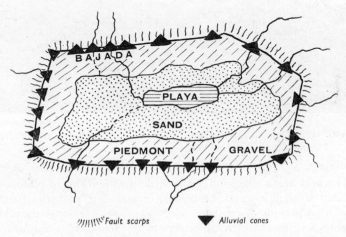

 Fault scarps   ▼ Alluvial cones

FIG. 118.—An Intermontane Basin

course, resulting from evaporation. Very often the coarser gravels of the piedmont zone have a water table, and this may emerge at the surface or be tapped at the inner margin of the zone.

Sometimes the alluvial fans or cones may coalesce along the foot of the scarp and create what is known as a bajada, which has a relatively gently inclined surface as compared with the slopes of the isolated fans (see Fig. 118). On the other hand, the base of the scarp may pass into a pediment of solid rock which slopes out gradually and passes under the alluvial deposits of the inner part of the basin. This pediment probably represents a surface that has been developed by the recession of the fault scarp through weathering and wind erosion proceeding under conditions of aridity, and later this surface has been overlapped in part by transported alluvial material.

Literary Services Mondiale Ltd.

FIG. 119.—Playa in Death Valley, California, United States of America
In the background is the piedmont gravel zone forming a bajada.

Inselbergs

The arid areas which have the structural features of plateaus, as distinct from intermontane basins, retain their plateau character after long periods of erosion, since very extensive areas have fairly level rock surfaces which are broken by residual hills, standing above the surrounding plains, whose tops represent an earlier plateau level. Such residual hills may show peculiar forms, and they are known in Africa as inselbergs, from the name given to them originally by the German geographer, Passarge. The typical inselberg is steep-sided and either round-topped or flat-topped. There is a most characteristic change of slope at the base of the steep sides, where a concave slope passes out into a rock pediment (Fig. 120). There is still

FIG. 120.—Inselbergs

much controversy over the way in which these inselbergs have been formed, but they are probably remnants of a surface, originally plane and level, which has been lowered by erosion under arid conditions. They may therefore be regarded as monadnocks (see p. 76). The concave basal slope has suggested to many investigators that water erosion has played its part in the development of these hills.

Cycle of Erosion in Arid Areas

Without attempting any detailed analysis of the probable development of a cycle of erosion, it would appear that the most predominant feature throughout such a cycle is the plateau surface, although it has been doubted whether deflation can lower surfaces to a very great extent unless it is aided by other erosional processes. The probable course of a cycle of erosion in intermontane basin areas is briefly as follows. The infilling of the basins and elevation of their floor levels by material derived from the adjacent uplands is the first stage. Each basin is separate in the early part of the cycle, but with the dissection of the intervening ranges some opportunity arises for the higher basins to drain to lower basins by means of spillways cut by water. The lower basins are then rapidly infilled, and reach a stage of maturity. Repetition of this continued infilling in higher basins, together with the reduction of the remaining areas of higher land, results in the final establishment of a rock pediment overlain by sand and sand dunes. Isolated remnants of the high terrain which originally enclosed the basins probably have the appearance of inselbergs. Thus whether the desert region was originally of the intermontane basin type or of the plateau type, the product of prolonged erosion is a plain or a plateau with upstanding highland remnants.

Loess

Under strong and steady winds, it is obvious that some of the product of desert deflation is removed far beyond the scene of its origin. It has been possible to record the extent of the removal of dust and sand by strong winds during great storms. In the United States of America it has been observed that wind has transported material of the soil from Texas to the environs of the Great Lakes. This wind-borne dust, after travelling hundreds of miles, arrives in a more humid region and rain may wash it down from the air to ground level, where it is trapped by vegetation. Such accumulations form loess and are found in steppe regions, marginal to deserts, and also in fairly close association with northern areas that were marginal to the Pleistocene ice sheets. Loess forms an extensive, yet discontinuous, belt across Eurasia from France through Central Germany, Russia, and Turkestan, and ends in the Shansi and Shensi plateaus of China, where it is thickest. In North America it is coincident with parts of the lava plateaus of Washington and Oregon and occupies a fairly continuous zone from Nebraska through the lower Missouri valley and along much of the course of the Mississippi below St. Louis.

Most of these deposits have been derived from glacial deposits, although in the Chinese area it is of desert origin. In all cases the loess shows the common characteristics of having been wind-borne, and the dust, which is yellow, has settled more often than not upon grassland areas. Despite the fact that stratification is absent, there is a columnar structure which, it is thought, is derived from buried grass roots that have decayed and produced fine vertical tubes, lined with calcium carbonate. These vertical tubes have imparted strength to these deposits, so that whilst loess may be eroded the material that remains stands up in the form of almost vertical walls. In China the deposits are so thick that they have been worn into deep 'bad land' areas. Much of this loess provides the alluvium that is distributed by the Hwang-ho floods in the Great Plain of Northern China. Elsewhere loess is less thick than in China and, although sometimes it forms low plateaus, it is usually found in plain lands. Whilst, from a geological standpoint, loess is a rock, it is readily converted into a loessic soil.

River Valleys in Arid Areas

Finally, reference must be made to the courses of rivers whose sources are in more humid areas, but which cross deserts or lands that are otherwise too arid to support perpetual river flow. The rivers are able to do little more than flow through these dry zones, and they cannot rely on the help of water from side streams for normal valley development. Thus the valleys are mainly trenched into the surrounding land surface. Down-cutting, where the base level permits, is of supreme importance. The Colorado River has, undoubtedly, produced the most spectacular course of all the great rivers which cross arid regions. In this case the river flows in places over a mile below the plateau surface in a canyon or gorge which is of varying width. The geological history of the region is quite complicated and, in one sense, this canyon represents the effects of rejuvenation with its incised river course, but aridity is responsible for such features as the sides where differential erosion of horizontal strata of varying resistance has created the fantastic scenery. Side streams are few in number, and there has been little or no reduction of the valley sides other than where spurs have been truncated or, in places, isolated, so that they stand out as solitary pinnacles of rock.

CHAPTER VIII

LIMESTONE AND CHALK AREAS

Character of Limestone and Chalk Rock

Limestone and chalk areas have their own markedly characteristic landscape. Such areas are usually upstanding in relation to adjacent regions, although neither rock is as resistant as granite or many sedimentary rocks. The upstanding relief is related to the pervious nature of the rocks and the effects of chemical solution upon them.

It is important to distinguish between porous and pervious rocks. A porous rock has many spaces which may become water-logged, as in the case of a clay or a sandstone, and, for that matter, chalk. This does not mean that water necessarily moves freely in these rocks, for often it is only along joint planes in chalk rock that water is readily obtained. Pervious rocks are open-grained or well-jointed, and many igneous rocks are pervious for the latter reason, but in limestone joints are so numerous that they produce land forms with unusual features. Many limestones are massive, in that normal stratification is absent, the rock having become self-cemented, so that it is but slightly porous, and are usually fairly resistant to weathering other than by solution along the joints. Thus most running water percolates down and along joint planes, so that subsurface erosion is the rule, whereas the surface remains dry and is not readily reduced by weathering and erosion. River water is derived from rain-water that contains carbon-dioxide in solution, and this permits the chemical solution of limestones which are otherwise insoluble rocks. Limestone (and chalk) may be regarded as insoluble calcium carbonate, but this compound is converted by rain-water into calcium bicarbonate which is soluble.[1]

Chemical studies of the soil have shown that there is a much higher concentration of carbon-dioxide in the soil than in the atmosphere. It is therefore probable that a considerable solution of limestone is effected by the soil moisture and humic acids of the soil in lieu of running water in underground streams. It is possible to see, as in the Buxton area of Derbyshire, the limestone surface after the thin soil layer has been stripped off, preparatory to quarrying. The

$$CaCO_3 + H_2O + CO_2 \rightleftharpoons Ca(HCO_3)_2.$$

surface thus exposed shows the characteristic clint formation which will be referred to later in this chapter.

The process of solution is very important, since it enlarges the spaces between the many joints and fissures of these rocks, and may encourage the formation of caves, although the latter are not very common in chalk areas. Both limestone and chalk hold a very large amount of water, and it can be shown, as in Fig. 121, that there is a water table, as marked by the dotted line, in a chalk escarpment. How-

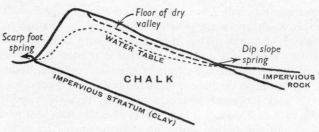

Fig. 121.—Water Table and Springs in a Chalk Escarpment

ever, it must not be assumed that the water can be obtained by sinking a well deep enough below the surface at any point. Water is enclosed in small pore spaces in the rock, and cannot move freely (otherwise it would produce a level water table), and some wells remain dry because they do not happen to strike a joint or fissure, and until a horizontal heading is made to connect with a water-bearing joint, there will be no flow in the well.

Karst Landscape

The land forms of limestone regions have, in all probability, been more thoroughly studied in Britain than has that of the chalk areas, while the most closely studied limestone region of all in the world is that of the Karst of Yugoslavia. Other areas of note which exhibit similar, if not identical, features to the Karst are parts of the Central Pennine region of England, the Causses in the Central Plateau of France, and several American areas, including parts of the limestone plateaus of the Appalachian Mountains in Kentucky and Tennessee, and also of the low-lying peninsulas of Florida and Yucatan. The Karst of Yugoslavia is a rather exceptional region of limestone, because surface drainage is largely absent, whereas in the other regions which have been mentioned it is usual to find some surface flow of water. Such streams are, however, almost invariably supplied from rainfall in neighbouring areas having impermeable rocks. Let us examine the chief features of the Karst landscape, many of which are common to all limestone regions.

Clints and Caverns

In many limestone areas the surface of bare rock may be fretted into a fantastic pattern of narrow clefts with intervening ridges and pinnacles of rock whose dimensions may be of the order of two or three feet. These are known respectively as grykes and clints in the Pennines. There are equivalent words for the same features in other countries, such as lapiés in France and bogaz in Yugoslavia. A very fine exposure of such a surface is to be

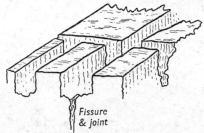

FIG. 122.—Rectilinear Pattern of Clints

seen above Malham Cove in Yorkshire. The fretting is almost entirely produced by solution along joint planes and an almost rectilinear pattern may be discerned, as shown in Figs. 122 and 123.

J. N. Kellett

FIG. 123.—Limestone Platform, near Galway, Eire

Clints are seen in the foreground and jointing of the limestone in the background. In the centre is a large round erratic of granite. Note that the sparse vegetation resembles that of the Mediterranean garigue.

Swallow holes, which are relatively large funnel-shaped depressions into which a surface stream may pass, are very common too. These are known in Yugoslavia as dolines, and as creux and by other names in France. They are also found very frequently as quite dry hollows where surface drainage has ceased after a period of normal river flow. Swallow holes are very common in chalk-lands. In some areas such holes have coalesced by individual growth to form quite extensive depressions known in Yugoslavia as uvalas. In some instances a swallow hole, or more usually a group of these inlets, will lead to a cavern below. The joint or joints below a swallow hole are enlarged by further solution, and where streams persist in their flow, there is a large amount of chemical solution which results in the opening up of a cavern (Fig. 124). Limestone caverns may be regarded as a series

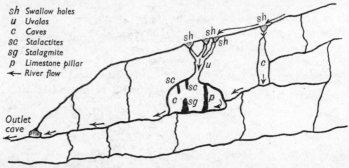

sh Swallow holes
u   Uvalas
c   Caves
sc Stalactites
sg Stalagmite
p   Limestone pillar
←   River flow

Outlet
cave

FIG. 124.—The Solution of Limestone

of connected chambers, some in a vertical plane while others are developed horizontally. It is not unusual for a cavern to show the spectacular development of stalactites and stalagmites. The process of solution shown above is reversible, and calcium bicarbonate yields calcium carbonate by chemical dissociation with the loss of carbon-dioxide to the air.[1]

Very slowly but surely the dripping water will form great masses of new limestones. Some are pendant from the roof of the caverns, called stalactites, while similar masses, known as stalagmites, are built up from the cavern floor. It is possible for a downward-growing stalactite and upward-growing stalagmite to join and give a pillar of limestone. Such deposition of calcium carbonate may be observed in chalk zones, where it occurs as calc-tufa at the margins of springs, issuing from a chalk escarpment. Nevertheless, much of the dissolved carbonate (or bicarbonate) is removed by river flow to the sea.

$$Ca(HCO_3)_2 \rightleftharpoons H_2O + CO_2 + CaCO_3.$$

Poljes

The development of uvalas is probably carried a stage further in the poljes (or polyen) which are found in the Karst. The poljes of Yugoslavia are very large depressions, and although solution has played a part in their formation, it is highly probable that they are partially tectonic in origin. Quite clearly in the late stages of Karst formation there has been a collapse of the cavern roofs, and this

Dorien Leigh Ltd.

FIG. 125.—Lake Scutari, Yugoslavia
A polje is seen with residual hills of limestone

has resulted in the growth of the depressions to the size which is typical of the poljes. The inner parts of these depressions are usually occupied by a group of residual hillocks of limestone which, no doubt, represent remnants of the cavern walls (Fig. 125). Although there are some areas in England which appear to have uvalas, it would seem that the poljes are almost exclusively Yugoslavian features. It is not at all certain that truly Karst landscape is present in France, nor, for that matter, in any of the other limestone areas which have been mentioned.

Dry Valleys

Some of the valleys in limestone country are dry, in that they have no surface flow of water in them today. The explanation of the cause of these dry valleys is simplified if it is realized that in the early stages of the development of a Karst landscape there was actually much water flowing on the surface. Until considerable opening up of joints had occurred, there must have been normal valley sculpture in such regions. This would proceed until the mass of limestone rock was sufficiently ramified with a network of interlaced underground channels. At this stage it seems likely that the streams which had

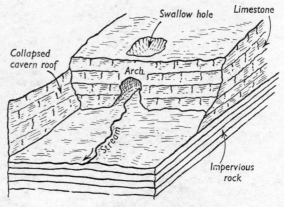

FIG. 126.—A De-roofed Cavern

cut valleys would plunge underground by way of swallow holes and leave unoccupied valleys below the points where they disappeared. The valley above Malham Cove in Yorkshire is of such origin (see Fig. 127).

With the development of caverns it is possible for a second type of valley to arise by the collapse of the cavern roof. A de-roofed cavern may expose a stream on its floor, but it is also possible, where the water table in the limestone is lower than the level of the cavern floor, for such an exposed valley to be dry. Not infrequently natural arches or bridges bestride the valleys at certain places. The latter are obviously sections of the cavern roof which have not suffered the general collapse (Fig. 126).

Many limestone areas have rivers which pass through very spectacular gorges, as for instance those followed by the Rivers Tarn and Lot in the Causses region of the Central Plateau of France. These rivers have apparently maintained normal erosion, whilst other features of limestone solution were being developed in the region, because they are fed from outside the limestone areas, and have thus

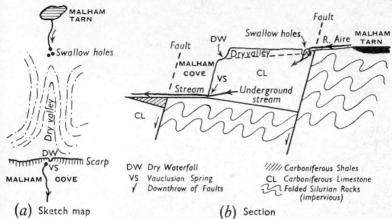

(a) Sketch map   (b) Section

DW Dry Waterfall
VS Vauclusian Spring
ⱱ Downthrow of Faults

///// Carboniferous Shales
CL Carboniferous Limestone
〰 Folded Silurian Rocks
     (impervious)

Fig. 127.—Malham District in the Limestone Region of the Central Pennine Uplands

received sufficient water supply to continue as surface flows which have cut down in the limestone and produced the narrow gorges.

Vauclusian Springs

The base of the limestone formation of rock is frequently marked by an impervious layer, since most other rocks must be relatively impervious in comparison with limestone. Thus where valleys have been trenched to the level of an impervious rock, it is common to find underground water issuing as springs from the limestone into these valleys. Such issues of water are known as Vauclusian springs. The spring which comes out at the foot of Malham Cove is of this type, although its formation is also due, in part, to faulting. The amphitheatre of limestone at Malham Cove marks a fault scarp which has receded, for over this, at one time, there was normal river flow. The course of this flow is clearly shown by the dry valley which stretches from the top of the Cove towards Malham Tarn, and also by the dry waterfall which can be seen most distinctly at one point as a notch on the edge of this scarp. The base of the scarp is in juxtaposition to impervious Silurian rocks, and it is here that the Vauclusian spring is to be found. In the dry valley above the Cove are swallow holes into which water may be seen going underground today, near another fault plane. It would seem that this water issues in the spring at Malham Cove, but this is not the case.

The Vauclusian spring is supplied from another part of the limestone area (Fig. 127). The main Aire drainage which arises in Malham Tarn issues in springs farther down the valley to the south of the Cove. In passing it may be noted that farther west the fault scarp produces the significant limestone edge known as Giggleswick Scar.

Cycle of Erosion in Limestone Regions

It is possible to review very briefly a sequence of erosion in limestone terrain as follows:

1. In the earliest stages surface drainage is the rule, and the opening of the limestone along the joints, preparatory to underground drainage.

2. Maturity is marked by the almost complete cessation of surface-water flow.

3. The late stages are indicated by the development of uvalas and poljes. In the Karst this stage has been reached, but it is questionable whether the cycle of erosion is so far advanced in any other limestone region. In the Yugoslavian poljes there are periodical floodings of the floors, which show that the water table is not far below the present surface. Consequently surface drainage might easily be restored as a permanent feature if there was a very slight elevation of this water table or, on the other hand, if there was a very slight lowering of the general surface level of the floors of the poljes.

Some limestone areas do not show any of these varied features. In some cases it is probable that they have been protected from such erosion until quite recently by the cover of an impervious rock. On the other hand, limestones may be less soluble, as in the case of magnesian limestone or dolomite, or the limestone layers may be interbedded with impervious strata which effectively prevent any very extensive solution. At Ingleborough and Pen-y-Ghent in Yorkshire cappings of horizontally bedded Millstone Grit, which are remnants of an erstwhile cover to the Carboniferous Limestone, protect the limestone from solution, so that it stands up in the form of flat-topped, mesa-like uplands above the general level of the surrounding limestone plateaus.

Reef knolls are masses of limestone forming isolated hills, which owe their relief to the resistant nature of the rock. Their origin is debated, but they are formed of very fossiliferous limestone. Clitheroe Castle stands on one of these knolls.

Chalk Areas and their Dry Valleys

Chalk land forms are more subdued in comparison with those of most limestone areas. The gorges and the abrupt, steeply inclined edges of limestone country are replaced by more open valleys and by rounded slopes, which are both convex and concave (Fig. 144). There is an absence of fretted surfaces (clints) and caverns in chalk, which is less well jointed than limestone, although swallow holes are fairly common, both on the upland surfaces and in the valley floors. Nevertheless, there is a widespread network of dry valleys, particularly on the dip slopes of chalk escarpments, and apart from areas of chalk upland which possess a cover of glacial deposits, there are very few streams. The dry valleys may be fed temporarily in the

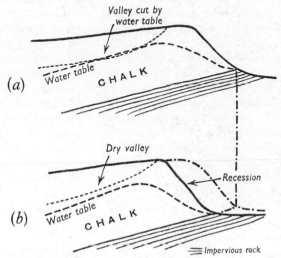

FIG. 128.—The Development of a Dry Valley by Recession of a Chalk Scarp

winter, today, by streams which are known as winterbournes. The word 'today' is used advisedly, since these valleys are evidently the products of previous water erosion. As shown in Fig. 121, the chalk uplands possess a water table which, in general, is below the surface of the region; where this table cuts the surface there will be an issue of water in the form of a spring. The impoverished drainage is clearly related to a depression of the water table, and stream flow will be renewed in the dry valleys only after an elevation of this table.

One explanation of the dry valleys correlates the fall in level of the water table with a decrease of rainfall in recent geological times. As there is no substantial evidence of such a decrease in rainfall, this explanation is unsatisfactory.

Another argument suggests that in the Pleistocene Ice Age the interstitial water in the chalk, lying just south of the ice sheets, was frozen, thus making the rock impervious and promoting surface drainage and the cutting of the valleys. Upon the return of a milder climate the chalk rock was rendered pervious once again and the drainage became subterranean. Although acceptable in the case of dry valleys in the chalk areas of Britain, this explanation cannot hold for similar, if not identical, valleys in parts of France which were too far south of the ice sheets to be affected in this way.

The third account of dry valleys is rather more complex than the others. The water table in the chalk is assumed to stand at a fairly constant height above the junction of the impervious stratum (usually a clay) at the base of an escarpment. This junction is marked by springs which are fed by water in the chalk. Most escarpments possess a series of such scarp-foot springs (see Fig. 121) which forms the spring line. In Fig. 128 two stages of escarpment development are shown; in (b) the scarp has receded from the position it held in (a). With the recession of the scarp, the level of the lowland at its foot has been lowered relative to the surface of the chalk at the crest of the scarp. Thus in time the valleys on the dip slopes of the escarpments must become dry downstream from their heads. It will be noted that this explanation calls for no decrease in rainfall, and it accounts for the formation of dry valleys in both glaciated and unglaciated regions of chalk.

CHAPTER IX

SHORE FEATURES AND COASTLINES—
CORAL REEFS AND ISLANDS

Introduction

In discussing the action of the sea as an erosive agent it must be realized that the zone of its action is the shore. It is, however, impossible to explain the intricacies of a shoreline without reference to both the geological structure and the history of the coast on which the discussion is centred. Briefly, much depends upon whether the coast (using the term for the margin of the land as distinct from the shore, where the action of the sea is operative) has suffered submergence or emergence (or both alternately) within recent geological times. The work of the sea is both destructive and constructive, and the former may be likened to degradation and the latter to aggradation by running water. In both cases the waves of the sea are of prime importance.

Wave Action

In the open ocean waves may be regarded as oscillatory movements of the water particles, in that the surface of the water body is deformed into troughs and crests. Unless the waves are driven by the wind (which is, of course, very usual) or moved by currents, there

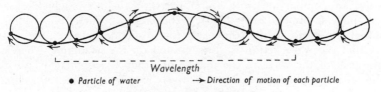

Wavelength

● *Particle of water* ⟶ *Direction of motion of each particle*

FIG. 129.—Wave Motion

is no forward movement of the water as a whole, because the water particles move in circular paths lying in a vertical plane (Fig. 129). The distance between adjacent crests (or between adjacent troughs) is known as the wave-length.

As waves enter shallow water offshore from land the effect of the decreasing depth of the sea is most pronounced. The paths of the water particles become elliptical, the wave-length decreases, the height of each wave increases, and eventually its crest overrides its base and the wave breaks forward, rushing up the beach as the swash or send. The momentum of this mass of water is in itself sufficient to create a force which has a shattering effect on obstacles lying on the seashore, whilst in caves in cliffs the alternating compression and expansion of enclosed air, trapped by the breaking waves, must cause much loosening of the rocks.

The swash is followed by the backwash. In waves of very short wave-length the swash is followed by a strong backwash which removes a great quantity of beach deposits. On the other hand, waves of long wave-length may have a very effective swash on breaking which carries deposits well up the beach, and these are not removed in bulk, as the backwash is relatively weak. The former waves may be regarded as destructive and the latter as constructive in action.

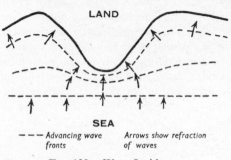

LAND

SEA

– – – Advancing wave fronts   Arrows show refraction of waves

FIG. 130.—Wave Incidence

However, like the wind, the water in a breaking wave is most effective as an agent of erosion when it is armed with rock fragments, and the resulting abrasive action is, over a long period of time, more important than the shattering effect. Wave fronts tend to strike the shore normally at all points. Shallow water, lying off a headland, retards the wave in comparison with those parts of the wave which enter far into adjoining bays before retardation occurs. This differential retardation causes the wave front to bend (or to be refracted) so that the wave action is concentrated on the headland, whereas the part of the wave in the middle of a bay is lengthened and its energy is thus dissipated over a relatively long front (Fig. 130). Thus the headlands are attacked more effectively by wave action, so that they are reduced and the shore is straightened in line with the heads of the bays and not further crenulated, as is often thought. Differential erosion by the sea on rocks of varying resistance may permit irregularities to persist locally on shorelines.

Whilst some parts of a shore are subject to the destructive action of the sea, other parts are being modified by the construction of sand spits and offshore bars. The formation of these has usually been attributed to the action of sea currents and to tidal influences, but, today, it is wave action that is regarded of primary importance in such constructions. One very important effect resulting from the breaking of waves on a shore is the production of longshore drift. Although it has been stated that waves tend to advance normally to the shore (Fig. 130), this clearly is not possible at all points and, as has been mentioned, a strong wind may give the waves a component which is oblique to the general trend of a particular shoreline. Nevertheless, the retreat of water in the backwash of a wave is always normal to the shore. It will be seen from Fig. 131 that material may advance on a zigzag course along the shore, as each successive incoming wave meets a retreating backwash with its load of fragments.[1]

Most sand spits have a number of lateral, spur-like ridges of deposited material on their landward sides, and this suggests that on-

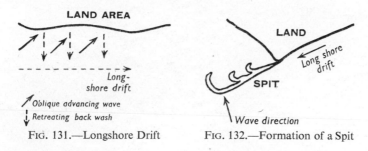

FIG. 131.—Longshore Drift   FIG. 132.—Formation of a Spit

shore waves, supplied with material from elsewhere by longshore drift, have built these prominences. Tidal influences are regarded as incapable of producing sufficiently large-scale deposition to create a spit, although they may move the smallest fragments. Spits are developed very often at a point where there is a change in the direction of the coastline, and they appear to be attached tangentially to the point (Fig. 132).

Dungeness and Morfa Harlech

Dungeness, on the Sussex coast, and Harlech Point, on the shores of Cardigan Bay, have been produced by a similar process. The beak-like Dungeness is at a position where effective wave action from both south and east has developed the two sides of the headland in front of the former shoreline, which is set back today behind

---

[1] The effectiveness of waves is determined also by the fetch or the length of water surface over which the wind blows. Briefly, a moderate wind blowing over a wide stretch of sea may be capable of creating larger waves than a strong wind with a smaller fetch.

Rye and Hythe, as well as being identified by a line of old cliffs on the northern side of Romney Marsh, some ten miles from the sea.

A significant change in the trend of the shoreline near Harlech has helped to create a spit which extends practically northwards, and behind which the present shore flats, known as Morfa Harlech, were constructed by wave action that carried material over the earlier-formed spit, by wind action producing dunes, and also by much alluvial infilling from rivers. The old cliffs, on which Harlech Castle stands, trend north-eastwards on the inner edge of Morfa Harlech (Fig. 133). This development has occurred mainly within historic

Judges Ltd.

Fig. 133.—Morfa Harlech, Merionethshire
This shows a constructed shoreline, looking north from near Harlech Castle.

times, since the castle at Harlech appears to have a water gate for entry and retreat by the sea, which is now at least half a mile from the cliff on which the castle stands.

The production of offshore bars is due to similar action, but in this case the bar is a line of deposits, which lies above water-level, and in its earliest stages, at least, is completely detached from the shoreline.

Submergence and Emergence

It is now necessary to examine the erosive effects of the sea in relation to coastal areas which have, in recent geological times, been either submerged or elevated. It is possible in each case to trace a cycle of erosion. Let us consider briefly what is meant by submergence and emergence. The level of the ocean depends in part upon the balance between the evaporation of water into the atmosphere and its return as rain, or more indirectly by rivers. Such fluctuations as occur, through the disturbance of such a balance, are normally very slight. However, in periods of great cold, such as in a glacial period, much water is locked up in continental ice sheets and the ocean level is lowered, although the burden of ice will probably depress the land masses by the operation of isostatic adjustment. Even so, it has been estimated that in the Pleistocene Ice Age the level of the ocean surface was 200–300 feet lower than its average level today. In post-glacial times the melting of the several ice caps has caused the level of the sea to rise, and transgressions of many tracts of land have occurred. Isostatic adjustments have, at the same time, compensated in part for this rise of the sea-level by the elevation of the land masses which has followed the unloading of the ice burden. To this day Scandinavia, as well as other northern areas, is rising very slightly year by year. Land sediments brought into the open ocean basins must displace water and raise the sea surface very gradually and thus help to produce submergence as well.

In addition, earth movements, such as result from faulting or the tilting of the land surfaces, may affect the level of both the continental blocks and the ocean floors and thus create local elevations and depressions giving similar results. It is perhaps sufficient to summarize these problems by noting that a rise in the level of the sea or a fall in the level of the land will produce submergence of a coastal area, whilst emergence of coastal areas is the result of a depression of the ocean or a rise of the land surface. On any particular coast the features which may be seen are the result, in the majority of cases, of the interaction of these several processes, and it is far from very easy to unravel the true sequence of events. Emergence may be identified through raised beaches, which may lie as much as a hundred feet above the level of the sea. Submergence may be detected in some areas, when at low tide the stumps of trees in submerged forests are visible just offshore.

Shorelines on Submerged Coasts

With the submergence of a coastal area the sea is brought into direct contact with relatively high land or former hill slopes. Wave action is restricted to only a few feet above and below sea-level, and the first development on a submerged shore is the cutting of a low cliff (Fig. 134 (*a*)). Then, with the aid of normal sub-aerial erosion, the part of the land edge above this low cliff is reduced. Thus the cliff recedes partly by wave cutting at its base and partly by rain wash from above. It may be noted that the dip of the rocks in relation to the shore is of some importance in regard to the rate of retrogression of the cliff as shown by Fig. 135. After a wave breaks there is a backwash, which transports the debris from the cliff and deposits

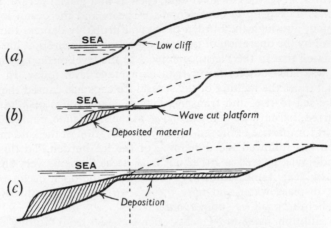

FIG. 134.—The Development of a Shore Zone after Submergence

it offshore (see Fig. 134 (b)) in the form of a submerged bank. Wave-cut platforms may be seen at the base of the cliffs, and give evidence of this process (Fig. 136). Further stages of this erosion show an extension out to sea of the bank of deposited material and the continued

FIG. 135.—The Effect of Dip on Shore Cliffs

The seaward dip in (b), combined with the jointing, will help to reduce the cliff more rapidly than in (a).

N.B. Wave cut nick at base of each cliff.

recession of the land margin, and there is a gradual approach to the form shown in Fig. 134 (c). This whole process may be regarded as a cycle of erosion, and the late stage corresponds to that of a graded river course. The recession of a cliff may result in minor streams having hanging valleys, as between the cliffs known as the Seven Sisters, near Beachy Head in Sussex.

FIG. 136.—Chalk Cliffs at Thornwick Bay, near Flamborough, Yorkshire

Note the wave-cut base of the cliffs, the wave-cut platform and the glacial deposits on the top of the Chalk.   The change in cliff slope marks the last.

Shorelines on Emergent Coasts

With the emergence of a coast the sea is, at first, in direct contact with a gently rising land surface, since the emergence has brought an offshore zone above sea-level. A line of low cliffs is produced by wave action, which may recede fairly rapidly, but the emergence has given a slight rejuvenation of the rivers in the coastal area, and in some cases valleys may hang at these low cliffs. Wave action, in reducing the land margins, collects material offshore in a series of bars (Fig. 82). The following stages of development are likely. Firstly, there is an infilling of the water zone between the inner shoreline and the outer bars, so that a new set of coastal flats is formed. These in turn are reduced to a line of cliffs in much the same way as has been described in the development of the cycle of erosion on a submerged shoreline.

Neutral Shorelines

It is not sufficient to refer to emergent and submergent shorelines. Some are neutral because there has been no recent change of level, and yet there is a development of the shore which has extended from the land as in great deltaic areas, alluvial plains, and around volcanic islands and, sometimes, coral reefs. Furthermore, many shorelines are compound, in that they have suffered both emergence and submergence within recent times, and have features resulting from both processes. The North Carolina shoreline is largely emergent in type, but some recent submergence has produced the features shown in the map, Fig. 137.

FIG. 137.—Shore Line near Cape Hatteras, U.S.A.

Types of Coastline

It is also necessary to discuss the coastlines, since they control some of the details of the shorelines. The relation of the coastline to adjacent land structures gives rise to various types of coast, of which two are known as Pacific and Atlantic.

Pacific coasts are so named because many thousands of miles of the coastline of the Pacific Ocean have a relationship with the fold mountain zones which surround this great basin (Figs. 45 and 46). The coastlines are broadly parallel to the fold trends of these mountains. In general, such coastlines show a remarkable tendency to be very regular or concordant, so that on the western shores of the American continents from Puget Sound in the state of Washington, to Puerto Montt in Southern Chile, there are few large inlets other than the harbour of San Francisco, with the Golden Gate, and the Gulf of Guayaquil.

Despite the fact that the coastlines of both British Columbia and Southern Chile are in concordance with the fold trends, there has been submergence which has resulted in the formation of archipelagos with the swamping by the sea of earlier valley floors and the lower slopes of ranges. Thus these coasts are very irregular.

The island festoons of South-east Asia are obviously related to the fold arcs of that region (see Fig. 45), although the separate islands form a discontinuous coastline. Submergence does not explain this broken coast, since there is evidence of recent uplift, and it has been suggested (see p. 41) that the region lies on the northern edge of a modern geosyncline and that these islands are literally the crests of uprising mountains.

Atlantic coasts are very different from Pacific coasts, since their main trend lines are oblique or at right angles to the structural trends of the adjacent land areas. Such coasts are said to be discordant and are usually very irregular, with alternating peninsulas and gulfs, as exemplified by the coasts of the Maritime Provinces of Canada and those of South-western Ireland. Nevertheless, some Atlantic coasts which are marginal to such continental or sub-continental plateaus as those of Eastern Brazil or of Western Africa are relatively regular. These may be called block coasts.

The terms Pacific and Atlantic, as applied to coasts, are somewhat misleading, since it is far from true to regard the coastlines of the Atlantic Ocean as being exclusively discordant any more than concordant coasts are peculiarly Pacific in their location. Pacific coastlines are evident in the Mediterranean Sea basin in Southern Spain (Sierra Nevada), Northern Africa (Atlas region), and Dalmatia (Dinaric Alps), whilst many of the coastlines of the Indian Ocean are Atlantic in nature, being marginal to such continental plateaus as those of East Africa, Western Australia, and to the sub-continental plateaus of Arabia and the Deccan of India.

Rias

We may now examine certain special types of coastline which owe their significant characteristics to submergence in that the regional structure is impressed upon the coastal pattern. Drowning of river-mouths by submergence creates the coastal inlets known as rias (Fig. 138). The word 'ria' is derived from the coast of Spanish Galicia where the westward-flowing rivers have submerged mouths, but the coasts of Brittany and South-west Ireland are very similar. Fig. 139 shows the coasts of Kerry and Cork, where the very straight bays, with their intervening peninsulas, are typical of the west. These coasts are related to a series of east–west upfolds and downfolds, the former giving parallel ridges and peninsulas and the latter carrying the main river valleys. It is the latter which have been submerged to produce the rias of Bantry Bay, Dingle Bay, etc. On the south coast is Cork Harbour, which is obviously more irregular than any one of the west coast rias. Here the mouths of several rivers have been drowned and a system of rias has been developed. Plymouth Sound, with its branches, is identical in origin.

Closely resembling these ria coasts are sections of the Dalmatian coast of the Adriatic Sea. Here, the totally submerged downfolds and the partially submerged upfolds of the land margins are represented by a series of parallel channels (or canali) and long narrow islets. Inlets resembling the ria systems of Plymouth and Cork are to be found farther from the open sea as in the Bocche di Cattaro. Here

V. C. Browne

FIG. 138.—Coast near Picton, South Island, New Zealand

The ridges belong to the fold mountain system at the north of the island. This is an example of a submerged coastline of the Atlantic type.

ridges of limestone (since the region is part of the Karst) are seen separating two arms of water which are linked by transverse narrow straits, where the upfolds have been breached. This is a Pacific coast, whereas rias are usually found on Atlantic coasts.

Fjords

The fjord inlets give yet another type of coast which owes much to submergence, but in this case there are other factors to consider.

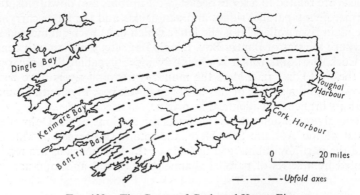

FIG. 139.—The Coasts of Cork and Kerry, Eire

Such coasts are found in fairly high latitudes; generally speaking, they are poleward of latitude 50° N. and S., and on the west sides of land areas. The coasts are usually fringed with islands representing the upper parts of partially submerged ridges or upland masses. The fjord inlets on the mainland of today have many branches, which upon close examination show a fairly rectilinear pattern. Fjords penetrate farther inland than do rias; some of the Norwegian examples have their heads as much as eighty miles from the sea.

Although the process of fjord formation has been debated, it is certain that all fjords have common features. These inlets are valley systems which have been affected by faulting and glacial erosion before submergence. The faulting is responsible for the rectilinear plan, whilst glaciation has developed U-shaped valleys with scalloped floors and tributary hanging valleys. Although the land on the sides of a ria may rise steeply from the water's edge, in a fjord the sides are almost vertical and shore margins are practically negligible (Fig. 140) other than at the heads of the main branches of the fjord, where infilling by rivers has produced deltaic areas of flat land. Furthermore, with a scalloped floor the depth of water varies from one part of the

Ewing Galloway, N.Y.

FIG. 140.—Naerofjord, Norway

The steep sides of the drowned U-shaped valley are shown.

fjord to another (Fig. 141), and in this respect a fjord differs from a ria, which becomes progressively more shallow as it is penetrated from the sea. In fact, with emergence, the fjord would become a lake-studded glacial valley.

Some Norwegian fjords have side branches, with shallow water compared with the depth in the main inlet. In such cases hanging

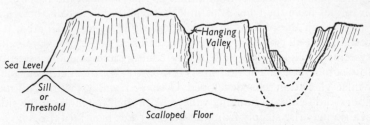

FIG. 141.—Section of Fjord Inlet

valleys have been involved in the submergence, and they are quite logically termed 'hanging fjords'.

Other Coasts

The flooding by the sea of land areas with subdued relief may give rise to inlets with many branches. Those on the East Anglian coast, formed by the mouths of the Rivers Orwell and Stour, are good examples. On the east coast of Denmark similar inlets give an irregular coast whose pattern is not unlike that of a fjord coast.

Locally, emergent coasts may have very complex outlines as shown in Fig. 137, where the development of offshore bars may enclose arms of the sea. The coast of the United States of America, in the neighbourhood of Cape Hatteras, is of this type. The process of infilling of such enclosed arms of the sea is clearly a deterrent to the establishment of useful ports.

Wherever such land-locked inlets of the ria, fjord, and other types have deep water penetrating far inland, maritime settlements and ports of major importance are likely to spring up. The value of the inlet of Port Jackson, at Sydney in Australia, lies in the abundant opportunities for wharfage on its many arms, while Plymouth, Milford Haven, and Brest are admirably suited as naval harbours. The fjord has, of all these harbours, most advantages, with its deep water and penetration inland (Fig. 142), but the position of all fjord coasts in high latitudes associated with cold climates has offset these advantages, since no major port of the world lies on such a coast, with the possible exceptions of those on Puget Sound, and Vancouver.

Canadian Pacific Railway Co.

FIG. 142.—Dyea Road, near Skagway, Alaska

An arm of a fjord on the Alaskan coast.

Local Coastal Features

The process of wave erosion in certain sections of a coast is partially balanced by deposition elsewhere as a result of the movement of eroded material by longshore drift. Thus the loss to the land by the destruction of the low cliffs of boulder clay along the coast of Holderness in Yorkshire is offset to a certain extent by the construction of Spurn Head. In Norfolk and Suffolk similar losses and gains are to be noted. Blakeney and Scolt Heads in the north and Orford Ness in the east are gaining on the whole in this region, while the Dunwich area is losing. Actually Orford Ness is shorter at the present day than it was in the past, owing to relatively recent changes in wave directions, so that there is active destruction of part of the long spit.

A spit may grow tangentially from some point on a coast, and eventually seal off a bay into which it has grown. The initiation of such a spit, as usual, is at a point where the coastline changes its direction. Maps of the Baltic Sea coast of Germany and Poland show this process in its various stages. Near Gdynia a spit, or neh-rung, is in an early stage of development, whereas farther east the Frisches Haff and the Kurisches Haff have been nearly sealed off, each by its own nehrung. In Devonshire, just north of Start Point at Slapton, there is an example of a coastal lagoon which has been produced by the complete sealing-up of Slapton Ley, an earlier bay, by a spit. On the French coasts of Aquitaine and Languedoc similar lagoons or étangs may be seen, and most deltas will afford numerous examples (see Fig. 67). The lagoon coast may be of value, since in Languedoc the course of the Canal du Midi threads some étangs. The Norfolk Broads are not dissimilar in origin, since they are the result of bar production by rivers in their shallow estuarine tracts.

When a spit joins the mainland to an island, there arises a tombolo. Chesil Bank, in Dorset, is an example of such a link bar or spit which joins Portland Isle to the mainland at Abbotsbury, some seven miles to the west of the island. Coastal flats may include rocky eminences marking former islands.

Lulworth Cove

Finally, a brief reference must be made to some very localized features resulting from the destructive action of the sea waves in conjunction with certain structural patterns. Rocks of varying resistance which outcrop on a coast, must create irregularities through differential erosion, and minor headlands, with their intervening bays, are thus produced, although, as described on p. 138, wave action will reduce such headlands with relative ease. Fault and strike lines usually control the sites of such headlands and bays (Fig. 10). Outcrops of rocks lying parallel to the coastline may result in the formation of an inlet known as a cove, as at Lulworth in Dorset, where the coast is bordered by an upfold of limestone with an outcrop of clay, backed by chalk hills, lying immediately inland. The diagram in Fig. 143 will show the geological details. Wave action has breached the outermost zone of limestone, and the sea has penetrated behind the cliffs and opened out, on the clay outcrop, the cove or land-locked bay which extends to the base of the chalk hills. To the east of Lulworth Cove this process has been carried a stage farther in Wor-

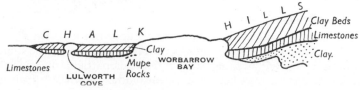

FIG. 143.—Part of the Coast of Dorset

barrow Bay, which has lost any cove-like pattern, but near Mupe Rocks there is a cove in the early stages of development (Fig. 144).

Stacks

Offshore from cliff coasts there are found stacks. Here joints or faults have given lines of attack from the sea. Undercutting of the cliffs along such lines of weakness produces, at first, coastal arches, but, later, these collapse and leave isolated pillars of rock known as stacks (Fig. 10).

Aerofilms Ltd.

FIG. 144.—Worbarrow Bay, Dorset

This bay lies east of Lulworth Cove. In the foreground is a minor cove, near Mupe Rocks. At the far side of the bay are two escarpments, the inner one of chalk, and the other of limestone (see Fig. 143). Dip slopes face inland. The chalk scarp shows the rounded nature of chalk landscape.

Coral Reefs and Islands

Of somewhat limited extent and significance in the world are the reefs and islands which are built of calcareous material whose source is almost exclusively organic, since it depends particularly upon the activity of corals, together with various other marine organisms. A coral is an organism known as a polyp; and has a soft, sac-like body which is normally embedded in a calcareous depression. The protective outer layers of calcium carbonate of these depressions are produced by the corals from sea water, which contains this substance in solution as calcium bicarbonate. Since corals usually live closely packed in colonies of thousands, the coralline structures of adjacent organisms become cemented together, often with the help of nullipores which are minute plants that attach themselves to the corals and secrete calcium carbonate as well. As corals die, so fresh corals build upon the structures of their predecessors, and the mass of coral rock is increased both laterally and very often, as will be seen, in an upward direction.

Corals are restricted to certain parts of the oceans since they require a mean annual temperature of 68° F., at least. Apart from local areas of the seas where the influence of warm currents of water permits them to thrive, they are not found far outside the tropics. Since within the tropics the temperature of the sea water is usually higher on the western sides of the oceans than in the eastern parts, there are most corals off the eastern shores of land areas. The abundance of corals in the Pacific waters off Queensland and the East Indies is therefore understandable. For similar reasons large numbers of corals occur in the Caribbean region, although the warmth of the Gulf Stream permits corals in Bermuda, at 32° N. Corals flourish in brackish waters, but the water must be free from the suspended terrigenous material that may be carried far out in the surface waters of the oceans lying off the mouths of large rivers. They grow best where breaking waves and tidal influences give them abundant supplies of fresh oxygen and marine food. Whilst corals can exist near the surface of the sea they must be covered with water during part of the day.

The coralline structures form fringing reefs to islands, barrier reefs that surround islands (with an intervening stretch of water known as a lagoon), and atolls which are like barrier reefs without an inner island. The atoll is essentially a ring-shaped island of coral rock enclosing a lagoon. Lagoons of this type are not the same as the coastal lagoons referred to on p. 150. The origin of these various features has given scope for much investigation.

The earliest theory of their origin, that of Darwin, called for a submerging island which was usually volcanic, since many reefs are associated with the Pacific islands of this type. The gradual development of the three types of reef is shown in Fig. 145. In (a) the island has a platform of coral or a fringing reef, perhaps half a mile in width, extending out from the shore at low tide, with its outer edge dropping steeply to the sea floor. In (b) the island has sunk, and the reef has grown upwards, keeping pace with the subsidence of the sea floor which it is assumed was not too rapid. The average rate at which coralline structures can grow upwards is about a foot in a

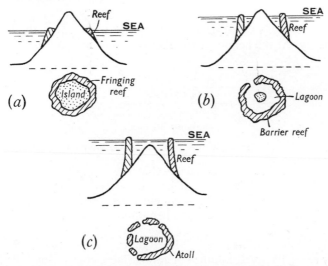

FIG. 145.—The Development of Coral Reefs and Atolls
(*After Darwin*)

decade. Too rapid a subsidence would cause the death of the corals and the reefs would cease to grow. Actually, corals can live at a depth of about 150 feet, and nullipores still deeper, but it is the rate of upgrowth which is of importance to this argument. The fringing reef of (a) has now become detached from the island, and is therefore known as a barrier reef which completely encircles the island, apart from a few breaks in its continuity. Further subsidence has caused the complete disappearance of the island above sea-level as in (c), but the reef has still grown upwards and produced an atoll.

There is one significant difference between the enclosed lagoons of the atolls which have been surveyed and the atoll as envisaged by Darwin's theory. Most lagoons, inside atolls, are uniformly deep over their areas instead of being deeper at their outer edges. The larger lagoons have a depth of some 250–300 feet, whilst the smaller ones are only about 150 feet deep. It is clear that sea waves are responsible for the comminution of dead coralline structures, and some of this waste is carried by waves over the reef and deposited in the lagoon. The quantity of coral mud which would be required to fill up the deep parts of the lagoon in Fig. 145 (*c*) and so produce a level floor around the subsiding island cannot be accounted for by the likely supply of such material by wave action alone. It would appear, from the extensive surveys that have been made in many Pacific atolls and their lagoons, that the reefs have been built upon a platform, generally at a depth of some 250–300 feet (Fig. 146). In order to explain this platform, Daly has put forward the following theory. At the height of the Pleistocene glaciation there was so

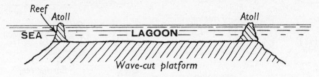

FIG. 146.—A Typical Pacific Atoll

much water held in the ice sheets that the level of the oceans was 200–300 feet lower than it is at present. When the sea-level was at its lowest, there were created wave-cut platforms at about this level. A low-lying volcanic island at the beginning of the Ice Age would be truncated by wave action as the slow depression of the sea surface occurred. Since the close of the Ice Age the sea has risen sufficiently slowly to enable fresh colonies of corals (most corals were killed off, even in tropical waters, by the cold of the Ice Age) to build up the bulk of the present-day reefs on these platforms. It will be observed that the development of atolls has also depended on submergence in Daly's theory, as in Darwin's, but the prime cause has been a rising sea-level, and not a sinking island.

The Great Barrier Reef, off the coast of Queensland, is somewhat different in that it does not surround islands but lies offshore from a line of small islands and therefore forms an extensive line of reefs for some thousand miles from north to south, broken at several points, some of which lie opposite to the mouths of the larger rivers of Queensland.

Part III

METEOROLOGY

CHAPTER X

INSOLATION AND TEMPERATURE

Introduction

Meteorology is a branch of science which may be justly called applied physics, since most of its measurements are exact, and from these an attempt is made to give an orderly statement of atmospheric conditions. Primarily, meteorology is concerned with the conditions which obtain in the lower layers of the atmosphere, some seven or eight miles in thickness. These are known as the troposphere, whereas the outer layers are known as the stratosphere. Regular measurements of weather conditions are made at meteorological stations which have been established throughout the world, although there are very large areas which are ill-equipped in this respect. Meteorological stations record the temperature of the air, atmospheric pressure and humidity, the force and direction of winds, the height, type, and movement of clouds, the amounts and types of precipitation (rain, hail, sleet, and snow), the extent of visibility, and many other factors such as the duration and intensity of sunshine, the temperature at the ground surface and below it at certain depths. To the geographer the most important measurements are those of temperature, pressure, winds, humidity, and precipitation.

Temperature

The temperature of the air is a measurement of basic importance. It represents the balance between the heat received from the sun and that dissipated from the surface of the earth by radiation and, to a less extent, by conduction, convection, and turbulence in the lower layers of the troposphere. Rays of heat emanate from the sun and pass through space and the atmosphere to the surface of the earth. It is when the rays strike the earth's surface that their energy is converted into heat. This supply of heat energy from the sun is known as insolation, but all the rays that come from the sun towards the earth do not succeed in reaching its surface. As the rays pass through the troposphere they are subject to scattering, absorption, and reflection by dust particles and clouds. The latter have a very potent influence in obstructing these rays, and many are reflected back into the atmosphere from their upper surfaces. Even some of

157

the rays which reach the earth are reflected from its surface, and so their energy is not converted into heat. However, a considerable fraction of the rays does produce heat on the surface of the earth, but the amount of heat developed on small unit areas of this surface depends on several factors.

Insolation and Incidence of Sun's Rays

The most important factor of all is the incidence of the rays on the surface of the earth since there is a great variation in the angle which the rays make with the surface. These variations depend mainly on the latitude and the season of the year. In Fig. 147 are shown three bands of the sun's rays which are equal in breadth, so that if we assume that there has been equal scattering and absorption in each

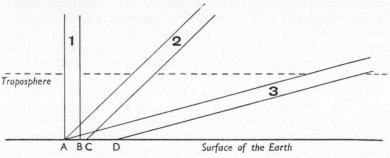

FIG. 147.—The Effect of Varying Incidence of the Sun's Rays on Insolation

band with its passage through the troposphere, then the base of each band may be regarded as producing the same amount of heat at the surface of the earth. As each band is inclined to the surface at a different angle, the area covered by each is represented by the lines AB, AC, and AD. Obviously the heat developed from the rays in band (1) is concentrated in a smaller area than either that of band (2) or band (3). The difference is dependent on the angle between the rays and the surface: rays at low angles cover greater areas than rays which are vertical or nearly vertical. Thus vertical rays, as in band (1), concentrate the heat received in each unit area, so that heating is at a maximum, whilst the rays at lower angles give relatively less heat to each unit area, reaching a minimum when the rays are horizontal. In the above statement it is assumed that the heating up has continued for the same period of time in each case. Actually it must not be assumed that the insolation received on each of the three surfaces AB, AC, and AD is the same. Clearly the rays at lower angles have had longer paths through the troposphere, so that the amount of absorption is greater than with vertical rays, and thus the amount of insolation received by each surface will be different.

Let us examine more closely the factors which cause the insolation to vary because of the variation of the angle between the rays from the sun and the surface of the earth. It may be assumed that all the rays of the sun which fall on the surface of the earth at a given instant are parallel to one another. In Fig. 148 the earth is shown at the position of its orbit when it is the summer solstice. Its axis is tilted at an angle of $66\frac{1}{2}°$ to the plane of the ecliptic[1], and the rays of the sun are parallel to this plane. Several rays are shown, striking the earth

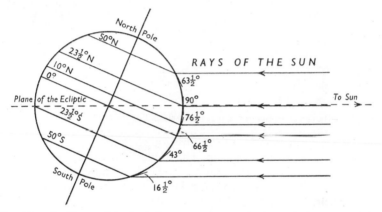

FIG. 148.—Incidence of the Sun's Rays at various Latitudes at the Summer Solstice

at various latitudes, with the angles between them and the surface of the earth numbered. Clearly the greatest insolation per unit area of surface is received at $23\frac{1}{2}°$ N., where the sun is overhead at noonday. At all other latitudes the rays are spread out in varying degrees according to the angle. Thus latitude has a control over insolation.

Seasonal Variation of Insolation

The angles given on the above diagram and on the two which follow are for the noonday sun. Hence the degree of insolation at any particular latitude as indicated by these diagrams is for the noonday period only. This observation is of importance, because allowance should be made for the number of hours of sunshine at different latitudes, and this has a great influence on the insolation

[1] This is the plane in which the orbit of the earth around the sun lies (it is the plane in which eclipses take place).

received during the daytime at any particular latitude and is not directly related to the angle of incidence of the noonday sun's rays.

In Fig. 149 the incidence of the rays of the sun at the same latitudes as given on the previous diagram is shown for the winter solstice,

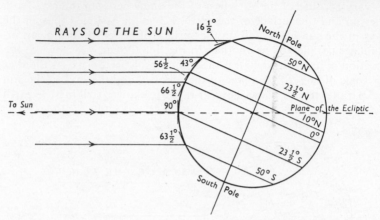

FIG. 149.—Incidence of the Sun's Rays at various Latitudes at the Winter Solstice

when the earth is at the opposite position to that of the summer solstice in its orbit around the sun. The angles between the rays of the noonday sun and the surface of the earth are all different from their previous values at corresponding latitudes, with the exception of the Equator. These differences may be shown in the form of a table:

Latitude	Angle between the rays of the sun at noonday and the surface of the earth at the summer solstice	Angle between the rays of the sun at noonday and the surface of the earth at the winter solstice	Difference
90° N.	23½°	0°†	23½°
70° N.	43½°	0°†	43½°
66½° N.	47°	0° *	47°
50° N.	63½°	16½°	47°
23½° N.	90°	43°	47°
20° N.	86½°	46½°	40°
10° N.	76½°	56½°	20°
0°	66½°	66½°	0°
20° S.	46½°	86½°	40°
23½° S.	43°	90°	47°
50° S.	16½°	63½°	47°
66½° S.	0° *	47°	47°
80° S.	0°†	33½°	33½°

* Rays are just horizontal.
† Rays are below the horizon.

In six cases there is a difference of 47° between the angle at one solstice and the angle at the other solstice. This is true of all places between the tropic and polar circle in each hemisphere. At all other times in the year the angles would show values between those given in the table above. It must be remembered that the solstitial positions are extreme ones, and that in the intervening periods of the year the sun is moving overhead between the tropics, where all places receive overhead sun on two occasions in each year. Thus at the various latitudes between and including $23\frac{1}{2}$° N. and S., shown in the table above, the noonday sun angle varies from 90° to the smaller solstitial value shown. Between the tropics and the polar circle ($66\frac{1}{2}$°) in each hemisphere this difference is 47° at all latitudes. But within the polar

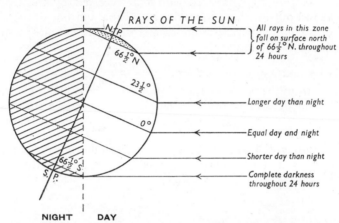

FIG. 150.—Varying Lengths of Night and Day at different Latitudes at the Summer Solstice (reverse is true at the Winter Solstice)

circles the difference is less than 47° and decreases polewards. Thus there is a seasonal variation of insolation which is related to both latitude and the revolution of the earth around the sun.

Within the tropics the noonday angles of the sun are never very low and their seasonal differences at any latitude are not so great as elsewhere. In middle and high latitudes when the noonday sun is at a low angle there is very great spreading of the rays at the earth's surface so that large areas receive an amount of radiation from the sun comparable to that which, in low latitudes, is concentrated on small areas. Thus, on these grounds alone, we may arrive at the conclusion that insolation, taken over the whole year, is greatest within the tropics.

Exclusive of other factors the larger the noonday angle the greater is the insolation received in the day. Reduced insolation, with rays at low angles, may be offset in very high latitudes in the summer by very long hours of daylight (Fig. 150). Whereas at the equator the lengths of night and day are the same throughout the year, north of $66\frac{1}{2}°$ N., at the time of the summer solstice, the day lasts for twenty-four hours. Thus there is the apparent paradox that the polar zones receive the highest computed insolation of any part of the earth's surface, but only for a very short period in the year is this true. As day and night are of equal length at the Equator throughout the year, and since the noonday angles of the sun's rays are never less than $66\frac{1}{2}°$, insolation in the equatorial zone is always great, but the regular diurnal development of a cloud cover helps to reduce the total insolation as compared with that received by the tropical regions just outside these zones. In the savana and hot desert areas the hot season temperatures are greater than those of any month at places near the equator.

Insolation and Aspect of Surface

Apart from the variations which have been mentioned, the resultant heating of any part of the surface of the earth may depend upon two other factors, namely, aspect of steep slopes and the nature of the surface.

A very steep hill or mountain slope in high latitudes may receive rays of the sun at a relatively high angle, so that the effective insola-

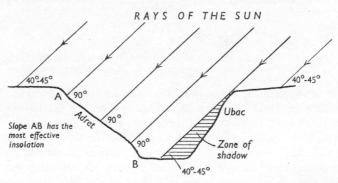

FIG. 151.—Incidence of the Noonday Rays of the Sun in Mountain Terrain

tion is much greater than on level valley floors or mountain tops in the vicinity (Fig. 151). It is quite apparent that in some Alpine areas the choice of the sites for small villages and their croplands is controlled by the aspect which affords the greatest insolation over the whole year. The slopes which are in the shadow for much of the day are usually shunned by these communities, and there are distinctive names for these shaded and sunny slopes (in the French Alps they are called 'ubac' and 'adret' respectively). Under overhead sun, the reverse would operate in deep mountain valleys in the tropics, and the high angle of the sun's rays would not give the most effective insolation on the sides of the valleys. The intensely cold slopes which are in the shadow during the day in the Andean areas may be quoted as examples.

Heating and Cooling of Land and Water

The nature of the surface on which the sun's rays fall also controls the heating up by day and, for that matter, the cooling down at night. For the moment we may regard the surface of the earth as being either of land or of water and compare these two types of surface. As important climatic distinctions depend upon the differences between these surfaces, it is desirable to examine fairly fully the factors which are involved:

There are three important factors:

1. The absorption of heat depends upon the nature of the surface. Water reflects the rays of the sun more than does land, and therefore for surfaces of equal area and with the same insolation water absorbs considerably less heat than does land in a given period of time, or, in other words, land absorbs more heat than does water. On the other hand, at night, when radiation occurs, land loses heat more readily than does water, because good absorbers are good radiators.

2. The heat capacity of the substance forming the surface on which the rays fall is of importance. Heat capacity is the amount of heat required to raise the temperature of a mass of a particular substance through one degree of temperature. Water has a much greater heat capacity than has land. Therefore, for equal masses and with the same insolation, water needs more heat than land to warm up to the same degree. In other words, land heats more quickly than water; it also cools more rapidly than water.

3. Finally, the rate at which heat is transmitted into the body of water or of land from the surface layers has its effect. As explained in (1), heat is absorbed by the surface layers of land and water at different rates. However, in water there is a great penetration of the heat rays, so that a slight warming of water occurs to fairly great depths. At night heat is brought to the surface by conduction in both land and water, but also by convection in the case of water, since once radiation has cooled its surface layers, these sink and displace slightly deeper and warmer layers which therefore rise to the sur-

face. The conduction produces a loss of heat from both water and land, as once the heat has reached the surface it will be radiated, in the case of land, very rapidly. But in addition to the slower radiation from water surfaces, there is a continual replenishment at the surface of fairly warm water from below by convection.

Thus, on all three counts, water warms up and cools down more slowly than does land, although another factor which helps to maintain the relative coolness of water is evaporation, since this causes heat to be lost from its surface layers.

This unequal heating and cooling is a daily phenomenon, but it can also be applied to the long-period contrast between summer and winter. Thus arises the distinction, that is so important in climatological discussions, between oceanic and continental climates, in which the difference between winter and summer temperatures is much less on the margins of the continents, near the oceans, than it is in situations within great land masses, particularly in temperate latitudes (see p. 354).

Yet another factor, which affects only the heating and cooling of land surfaces, is the type of soil. Soil contains air spaces and, since air is a poor conductor of heat, in very porous soils heat is concentrated in the surface layers which warm up readily by day, unless there is a considerable water content in the pore spaces, when the heating is slower and may eventually be carried to fairly great depths. A dry sandy soil will warm up in its surface layers quickly, whereas a close-textured clay soil with less air and a fair amount of water in the pore spaces will warm up slowly. On the other hand, the former soil cools more quickly than the latter. Surfaces of snow and ice may be compared with sand in so far as they lose heat rapidly, but they do not gain heat readily because the reflection of the incoming heat rays from these surfaces is very great.

From the foregoing arguments it would seem that at night the temperature should fall steadily until the renewed onset of insolation with daytime. This rarely occurs, because clouds act as reflecting surfaces for the radiant heat rays, which are therefore returned to the surface of the earth and so help to maintain the heat balance. Also the introduction of warm air into a given area often prevents the normal nocturnal chilling. In Britain importations of tropical air in winter are particularly effective in this respect.

Temperature Records

Air temperatures are measured at meteorological stations by thermometers which are housed in a Stevenson screen. This screen is a wooden box with ventilated or louvred sides which permit the free movement of air through the box and yet prevent any direct or reflected rays of the sun from entering it. The screen is mounted preferably over a grass surface so as to reduce the chance of increased heating such as would occur by the reflection of the sun's rays from bare soil or a roof top. For the measurement of air temperatures in the shade, that is, in the screen, two thermometers of special construction are used to give records of the highest (or maximum) and the lowest (or minimum) temperatures in a given period, irrespective of when the instruments are read. Obviously, unless some device is employed, it would be virtually impossible to determine the extreme temperature that each thermometer has actually recorded during a given period. It is usual to read the thermometers at frequent and regular intervals. Some meteorological stations take only one set of readings in every twenty-four hours, but it is more usual for several sets of readings to be made during the day. In addition to the maximum and minimum thermometers, the screen should house a self-recording instrument known as a thermograph. This gives a continuous trace on paper of the temperature by means of a pen which moves on an arm coupled to a coil of two metals. Changes in temperature are reflected by the expansion and contraction of this coil, and these movements are transmitted to the pen. The trace is known as a thermogram.[1] All temperatures used for geographical purposes in this country are given in degrees Fahrenheit.

From the highest and lowest temperatures recorded in a given day the mean daily temperature for that day is obtained as an average of the two values. The difference between these two temperatures is known as the daily range of temperature. For example, if the maximum and minimum temperatures on a certain day are 60° F. and 40° F. respectively, then the mean temperature will be 50° F., and the daily range will be 20° F. Over a month readings are kept for each day, and the average of the mean daily temperatures for the several days of the month yields a figure known as the mean monthly temperature. For climatological purposes the mean temperatures for each of the twelve months of the year are used for comparisons. However, the figures quoted in climatic statistics are based on long-period observations covering several years. The period that is considered adequate to include all the likely variations which may occur is one of thirty-five years. Thus a mean temperature for a given month is really an average of probably at least thirty-five mean

[1] Various meteorological instruments, described in this book, and others, are well illustrated in a filmstrip, 'Observing the Weather' by E. P. Boon, M.Sc., published by Rendell and Wilson, 176A Ashley Road, Altrincham, Cheshire.

monthly temperatures based on records extending over thirty-five consecutive years. With the more uniform conditions which prevail in the tropics as compared with the temperate areas, a ten-year period is often sufficient. The difference between the highest and lowest mean monthly temperature in the year gives the mean annual range of temperature.

It is of value to take into account the extremes of temperature, since they may be critical to plant growth. It is possible, from daily temperature records taken over many years, to extract the mean of the daily maxima and minima for a period such as a month. The two means which are thus obtained are known as the mean daily maximum and the mean daily minimum temperatures respectively, and their difference is the mean daily range of temperature. By taking the means of the highest and lowest temperature recordings in a particular month in each year over a period of observation, it is possible to obtain the mean monthly maximum and the mean monthly minimum temperatures, and their difference is called the mean monthly range of temperature. The absolute highest and lowest temperatures recorded over a given period are also of some value.

Lapse Rate of Temperature

Elevation produces a general fall in the temperature of the air at any latitude, the decrease being at the rate of approximately 1° F. for every 300 feet. This is known as the lapse rate of temperature in the troposphere. The following statistics will illustrate this lapse of temperature:

Fort William	January mean			July mean		
(171 feet)	temp.	38·7° F.	} 15·3° F.	temp.	57·1° F.	} 15·4° F.
Ben Nevis	January mean			July mean		
(4,406 feet)	temp.	23·4° F.		temp.	41·7° F.	

Note that the difference in elevation between Fort William and Ben Nevis is about 4,200 feet. With a lapse rate of 1° F. in 300 feet there should be a temperature difference between the two stations of about 14° F. The differences in January and July are in reasonable accord with this computed difference.

It is possible for the temperature to increase at high levels; in which case, the lapse rate of temperature in the troposphere decreases. This is known as a temperature inversion and the increase of temperature takes place at an inversion level.

Isothermal Maps

In order that a general comparison of temperatures in different parts of the world may be made, it is usual to make use of iso-

thermal maps. Isotherms are drawn as smooth lines with temperature values obtained from as many places as possible for a given
period, such as a month. They are boundary lines, and must be regarded as separating zones with temperatures above and below a
certain value. Reduced temperatures are usually employed for isotherm maps, when they are drawn for world distributions as shown
in most atlases. Since altitude produces lower temperatures than at
sea-level in all latitudes, it is of little value for comparative purposes
to draw isotherms from actual mean temperature values, since they
will give, in essence, a contour map with inverted values, i.e. low-
value isotherms will delimit elevated areas and vice versa. Reduction
of temperature is effected by adding to the mean temperature
1° F. for every 300 feet that the station of observation is above sea-

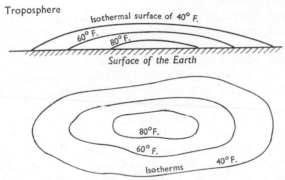

FIG. 152.—The Relation of Isothermal Surfaces to Isotherms

level. The map which is obtained, showing mean sea-level isotherms,
is of some value, although its worth is often over-estimated. However, for broad comparisons of the temperatures in widely separated
areas, it is the only map which has been devised. It should be clearly
understood that these maps have been produced by smoothing the
actual temperatures, and are, in many ways, theoretical maps, although they do combine the effects on temperature of such factors
as latitude and continentality, and are of great value in the study of
climatology.

It is possible to regard an isotherm as the intersection of an isothermal surface with the surface of the earth. Fig. 152 shows the
relationship between isotherms and isothermal surfaces.

World Distribution of Temperature

World maps of mean sea-level isotherms for the months of
January and July are given in most atlases. Study of such maps will

reveal that the zones of greatest heat, with mean monthly temperatures in excess of 80° F., are not only tropical, but extend to more than 40° N. in the northern summer and to about 30° S. in the southern summer. The two maps show clearly the migration of this hot zone, which is obviously related to the movement of the overhead sun between the tropics. However, the extensive land areas in the northern hemisphere give a much greater hot zone than in the south in the respective summer seasons, thus showing the effects of the great heating of land in summer. This difference between the hemispheres is further emphasized by the three fairly large areas in North America, North Africa, and Asia respectively, which have mean July temperatures of over 90° F. There are two corresponding areas in southern Africa and North-west Australia in January, but they are fairly small.

Polewards from these hot zones in each map there is a general decrease of temperature as would be expected. In the southern hemisphere the isotherms show a considerable uniformity of trend from east to west, which suggests that, in general, temperatures decrease regularly with increasing latitude. However, in the northern world the isotherms are far from regular in their courses across the maps. In winter there are marked poleward bends over the oceans and equatorward bends over the continents, indicating that along a given line of latitude the continents are much colder than the neighbouring oceanic areas. In January, along the Arctic Circle, it may be observed that the interiors of Canada and Siberia have mean temperatures of −20° F. and −30° F. to −50° F. respectively, whilst over the North Atlantic Ocean, between Norway and Iceland, the mean temperatures are between 30° F. and 40° F., giving a zone of winter warmth. The cold pole of the world is located on the Arctic Circle in Siberia, which in January is colder than the North Pole, whilst there are similar areas of intense cold in Greenland and northern Canada. The difference between the most northerly part of an isotherm loop over an ocean and its most southerly position over a continent in January in the north may be of the order of more than twenty degrees of latitude. The July map shows that in the northern hemisphere the contrast between continents and oceans is reversed. Observing the North Atlantic Ocean and the adjacent regions of North America and Eurasia the difference of temperature between land and sea at a given latitude is of the order of 10°–15° F. Both interior North America and Eurasia have summer temperatures of about 70° F. in the fifties and sixties of north latitude, whereas over the adjacent oceans the temperature is about 55°–60° F. This contrast between the summer and winter isotherms in the north is related to the difference in the heating and cooling of land and water as discussed on p. 163.

In the southern world the ocean surfaces are predominant, which explains the more uniform behaviour of the isotherms, although there are slight poleward loops over the hot continents in January (summer) which are maintained in July (winter), except in the case of the widest southern land mass, Australia, where there is an equatorward loop. In fact the loops in South America and Africa are produced mainly by warm and cool currents of water passing along the shores of the continents (see p. 348). The equability of temperature in the southern continents is shown by the fact that the differences between mean summer and winter temperatures are of the order of 10° F. as compared with 70° F. and even 80° F. which are typical of the interiors of the northern continents.

Other Temperature Maps

These general differences are shown even more vividly by maps that indicate the mean annual range of temperature over the world. On such maps the very small range of 5° F. which is restricted to an equatorial belt is most striking. The effect of a land mass is seen by the great ranges in the northern continents, whilst the moderating influence of oceanic areas is particularly clear in the southern world. Perhaps the most outstanding feature of all is afforded by the extensive northward loop of the isopleth[1] of 20° F. mean annual range of temperature in the North Atlantic Ocean, showing the tempering effect of water.

Maps of temperature anomalies may also be found in some atlases. An anomaly of temperature is the departure of the mean sea-level temperature from the average temperature for the latitude of the place concerned. This latter temperature is a purely theoretical one.[2] Anomalies of temperature may be positive or negative, and they show where parts of the surface of the earth are respectively warmer or cooler at a given time than they would be if the earth had a uniform surface so that there was a regular decrease of temperature in a poleward direction.

It is possible to draw a line known as the thermal or heat equator. It joins points on each line of longitude on a world map which have the highest mean temperature for a given period. The annual heat equator is shown on the map (p. 276), and it indicates that the

---

[1] An isopleth is a line joining places with the same quantity. The term covers all such lines as isotherms, etc. (Gr. Isoplethes—equal in quantity or number.)
[2] This temperature is obtained from readings taken at as many places as possible, spaced evenly along a line of latitude.

zone of highest average temperature in the world tends to reside mainly in the northern hemisphere. Whilst the slightly longer sojourn of the overhead sun in the northern hemisphere may help to explain this fact, the more extensive land masses in the north play an important rôle.

CHAPTER XI

PRESSURE AND WINDS

Pressure

The various gases which make up the atmosphere have a very considerable weight that exerts a force upon the surface of the earth which is known as the air pressure. The pressure at ground level is equivalent to the effect of a weight of about 15 lb. on every square inch of the surface. This pressure is not appreciated by human beings,

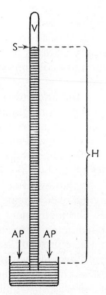

FIG. 153.—The Simple Mercury Barometer

AP = air pressure exerting a force on mercury in the reservoir.

H = height of mercury in tube which balances air pressure.

V = vacuum.

S = precise scale for measuring upper surface of mercury to at least 1/100 inch.

since our bodies are adapted by nature to withstand such weights. The density of the air in the lower layers of the atmosphere is greater than in the higher ones, which means that the weight of air or the pressure does not decrease uniformly with ascent.

Measurement of Pressure

Air pressure is measured by a barometer (Gr. *baros*—weight, heaviness). There are two types of instrument: one makes use of mercury and the other of a vacuum. In the simple mercury barometer, which is illustrated in Fig. 153, a vertical glass tube contains mercury and opens into a reservoir of mercury. Normally the air pressure (AP) supports a column of mercury thirty inches in height (H), and for this reason the tube must be about a yard in length. Above the mercury there is a vacuum (V). Since the pressure of the air is continually changing, the varying length of H is a direct measure of these changes. It is thus customary to indicate the air pressure in terms of inches or millimetres of mercury. In most parts of the world the fluctuations of pressure which occur over several hours, and even days, are usually extremely small, and it is necessary to use a very precise scale, which has a vernier, in order to register the changes in the height of the mercury column to the nearest hundredth of an inch. It is usual, however, to record the pressure of the air in millibars. The weight of the air may be expressed as a force of so many dynes. Pressure at sea-level is equivalent to a force that is usually in the neighbourhood of 1 million dynes per square centimetre, or 1 bar[1]. A millibar is therefore equal to a force of 1,000 dynes per square centimetre. When the pressure changes by 1 millibar, it is equivalent to a rise or fall in the mercury column of about $\frac{3}{100}$ inch.

The second type of barometer is known as the aneroid (without liquid), and it consists of a small metal box with corrugated sides, from which most of the air has been evacuated. Such a box will react very readily to changes of air pressure, since its sides will move in or out very slightly as the pressure increases or decreases respectively. Such slight movements that occur in the sides of the box are coupled with a series of levers and other mechanism, so that eventually these movements are recorded by a pointer moving over a scale which reads directly the value of the air pressure. The instrument is calibrated by reference to a standard mercury barometer. Obviously, its dependence on mechanism means that in the course of time the readings become inaccurate, so that frequent checking with a mercury barometer is essential. The compactness of the instrument is of great value, since it is easily carried about; precise instruments of this type, no larger than a pocket watch, may be constructed. It is also readily adapted to give a continuous record of pressure changes by a line traced by a pen moving up and down over paper on a revolving drum. This instrument is known as the barograph and the

[1] Standard pressure is regarded as that produced by a column of mercury 76 cm. in height. In a tube of 1 sq. cm. cross-section, this is equal to the weight of 76 c.c. of mercury. The density of mercury is 13·6, so that the weight of 76 c.c. of mercury is 13·6 × 76 grammes. One gramme weight is equivalent to 981 dynes. Therefore, standard pressure is equivalent to a force of 13·6 × 76 × 981, or 1,014,000 dynes per square centimetre, which is slightly more than 1 bar.

trace of pressure is a barogram. In British meteorological stations the Kew type of barometer is standard; it is very like Fortin's standard mercury barometer.

Variations of Pressure

In the first few thousand feet above sea-level pressure decreases at the rate of about 1 inch for every 1,000 feet. At about 17,000–18,000 feet the air pressure is reduced to about half of its sea-level value, but there is very much more than another 18,000 feet of atmosphere above this height; in fact, the atmosphere reaches to heights which are better expressed in miles rather than feet. In the case of readings by all barometers, the effect of temperature is of importance. Without giving a detailed explanation of how pressure readings are reduced to sea-level, it may be observed that a formula has been obtained to make this correction. It involves a knowledge of both the height and the temperature of the instrument. The force exerted by the mercury in a barometer varies with the force of gravity, whilst the density of the mercury varies with the temperature. Both the force of gravity and the temperature vary over the surface of the earth, and thus further corrections have to be applied to pressure readings in order to obtain values which are reduced to sea-level. Pressure readings are therefore corrected so as to correspond to readings made at sea-level at 45° latitude and at a temperature of 32° F. The force of gravity is considered to be at its mean value at 45° North or South, since it depends primarily upon the distance from the surface to the centre of the earth, and as the earth is not a perfect sphere but flattened at the poles, the force will obviously vary with latitude.

Under normal conditions the pressure of the air fluctuates fairly considerably, but it usually ranges between about 28·5 and 30·5 inches (or 965 and 1,033 millibars). Just as there is a daily range of temperature at places on the surface of the earth, so is there a diurnal rhythm of pressure. This rhythm is well marked on barograms at tropical stations, but in middle and high latitudes considerable fluctuations of pressure are experienced within short periods, often within a few hours, which make it very difficult to detect this rhythm. Nevertheless, it exists, and may be shown by averaging out long-period records for extra-tropical stations. The air pressure increases to a maximum about 10 a.m. and 10 p.m. and decreases to a minimum at about 4 p.m. and 4 a.m. Seasonal changes of pressure may be observed as well, and these may be closely related to seasonal temperature changes. For instance, in the interiors of the great northern continents the winter chill is reflected in the higher values of pressure, whilst the summer heat results in lower pressure than is usual in other areas of the world. The low-pressure zone of the equatorial belt, usually known as the Doldrums, is partly dependent on the uniformly high temperatures which prevail there.

High and Low Pressure

A knowledge of the exact values of the air pressure is not of compelling importance to the geographer. In general, the recognition of high- or low-pressure zones is sufficient, because it is the effects of pressure distributions that matter to us, and the interactions between areas of *relatively* high and *relatively* low pressure are of paramount importance. Again, between two areas of markedly high pressure there must of necessity be air at *relatively* lower pressure, although it may not be low as judged by the value of normal pressure. In studying the wind systems of the world the distribution of these *relatively* high- and low-pressure zones is of greater value than is the precise knowledge of average pressure values which obtain in different areas as shown by the seasonal pressure distributions in atlas maps.

Isobaric Maps

It is possible to draw isobars, which are lines joining places with the same pressure value for a given month or for any period which may be considered desirable. The values of pressure used are reduced to sea-level. Reduced isobars are perhaps of even less value than are reduced isotherms, because pressure changes at ground-level are often dependent upon changes at high levels in the atmosphere which cannot be indicated by sea-level isobars. However, sea-level isobars do present a picture of the average conditions of pressure over the earth's surface, although large areas on atlas maps are covered with isobars which have been drawn from the most scanty information. Again, the student is warned against placing too great a reliance on such maps, but to regard them as useful summaries of world-wide conditions. It is convenient to visualize isobaric surfaces which are comparable with isothermal surfaces as described on p. 167. From both isobars and isobaric surfaces it is possible to conceive of a pressure gradient, which means that there is a fall of pressure at right angles to the isobars and in a direction from high values of pressure to low values. This means also that there should be a movement of air across isobars according to the direction of the decline in pressure and also across isobaric surfaces in a horizontal direction as shown diagrammatically in Fig. 154.

World Distribution of Pressure

A glance at atlas maps of the world showing mean sea-level isobars for January and July will indicate that there are contrasts between these months and between the two hemispheres. In both maps
it will be seen that in the southern hemisphere the mean sea-level
isobars show a tendency to run from east to west, and the same trend
is noted in the subtropical ovals or cells of high pressure lying over
the oceans. Such behaviour is rare in the isobars of the northern
hemisphere, although the east–west trend is shown by the oceanic
high-pressure cells in the subtropical latitudes of the north. There
are five of these oceanic areas with high pressure in the world as
seen on maps, and they are permanent, although in winter in each

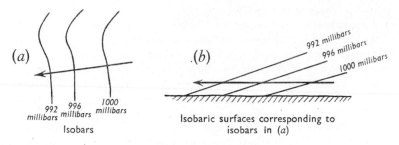

FIG. 154.—Pressure Gradient

hemisphere they are linked to one another by extensive areas of predominantly high pressure which lie over the intervening continents,
and in the north these extend polewards. Between the two sets of
subtropical high-pressure cells lies an equatorial belt of permanently
low-pressure air (the Doldrums) which shows a strong tendency to
extend polewards over the continents according to the hemisphere
which is experiencing summer. These extensions are instrumental in
restricting to oceanic areas the high-pressure cells referred to above.
The northern continents, particularly Eurasia, develop higher pressure in winter and lower pressure in summer than their southern
counterparts.

There are two well-defined low-pressure areas in winter over
the northern oceans. One is centred over the North Pacific Ocean
near the Aleutian Islands at about 55° N., whilst the other appears
to be centred at 60° N. over the North Atlantic Ocean between
Greenland and Iceland. In summer the former has disappeared
from the July map of pressure distributions, but there are vestiges

of the latter on this map. There are no counterparts to these low-pressure areas in the southern hemisphere; rather does one find a complete circumpolar belt of low pressure in both winter and summer which is marked by the very even trend of the isobars between 40° S. and 60° S. The values on the isobars show that the lowest pressure is probably aligned from east to west in a belt near the line of latitude of 60° S. Apart from a few scattered observations which have been made in Arctic and Antarctic areas, little is known of the pressure distributions near the two poles, but it is fairly evident that pressure is relatively high over the polar caps at all seasons. It is thought that high pressure prevails only in a shallow surface layer, about 6,000 feet thick, above which is low pressure.

These distributions in winter and summer reflect the continentality of the northern hemisphere and the predominant control of the ocean surfaces in the southern world. The winter chill of the northern land masses enhances the values of high pressure, so that, relatively, the high pressure of the subtropical cells is subdued. At the same time the more northerly oceanic low-pressure areas are thrown into greater relief by the development of the continental high pressure. The highest pressure (shown by the 1,040 millibar isobar) has its focus in Central Asia, where structural features induce very cold, heavy air to be concentrated in large mountain-girt basins, so that this region is quite exceptional.[1] In summer great heating of the interiors of the northern continents produces low pressure which causes a relative augmentation of the subtropical high-pressure cells and a relative, if not complete elimination of the oceanic low-pressure areas over the North Pacific and North Atlantic Oceans. In Asia the core of lowest pressure, with isobaric values of less than 995 millibars in July, is more southerly than the winter focus of high pressure, being located over the north-west of India. Summer low pressure in North America is in much the same area as the winter high-pressure zone.

The extensive ocean surfaces of the southern hemisphere remain fairly uniform throughout the year in regard to temperature, and this is reflected in the relatively simple pressure distributions of the south. The continental areas have but slight effect, and it is only in the summer (January) that there is a development of distinct continental low-pressure areas. These may be regarded as poleward extensions of the intertropical low pressure of the Doldrums belt and they subdivide the subtropical high-pressure region into its three separate oceanic cells.

---

[1] It should be remembered that the pressure values quoted are reduced ones and that actual values will be much less.

Wind

Wind is air in motion, and a pressure gradient must be established before a wind may blow. If a series of isobars are considered as in Fig. 154, then a wind should blow at right angles across them from high pressure to low pressure as shown by the arrow. A wind has both force and direction; the former is indicated by the spacing of the isobars, which may be regarded as showing steep or gentle pressure gradients in much the same way that closely or widely spaced contours show contrasted gradients on Ordnance Survey maps. The meteorologist classifies winds according to their speeds, in knots, which are recorded by a very complicated instrument known as the pressure-tube anemometer that gives a continuous tracing or anemogram. The wind has a strength which may be given in terms of speed or as a pressure (lb. per square inch). However, it is customary to use the Beaufort scale of wind force which was drawn up by Admiral Beaufort for use at sea in the days of sailing vessels. There are thirteen forces on this scale, and they are numbered from 0, 1, 2, . . . 12, ranging from calm (0) to hurricane (12) through various grades of breeze and gale. Originally the scale was developed on a purely qualitative basis according to the effects that the winds had on the sailing of ships, but in modern times it has been placed on a quantitative basis, using measured wind speeds. The velocities which are used range from less than 1 knot for calm to over 65 knots for a hurricane. The scale will be found in full in most standard textbooks of meteorology, and may be referred to by the reader.

The anemometer, referred to above, is an adaptation of the wind vane, and for most geographical purposes a knowledge of the direction of the wind is of greater importance than its force. A wind vane merely gives the direction of the wind; its arrow head points *into* the wind, and therefore indicates the point of the compass *from* which the wind is blowing. Consequently, a wind is named by the point of the compass indicated by the arrow.

Circulation on a Stationary Globe

Assuming for the moment that the earth is stationary, in that it does not rotate on its axis, then there would be a very simple system of air movement in the troposphere. There would arise, in fact, a transfer of heat by air currents from the hot equatorial zone to the cold poles in the way that is suggested by Fig. 155. Let us see how this movement may arise by the consideration of isobaric surfaces. In a perfectly uniform atmosphere which is resting on a uniform surface

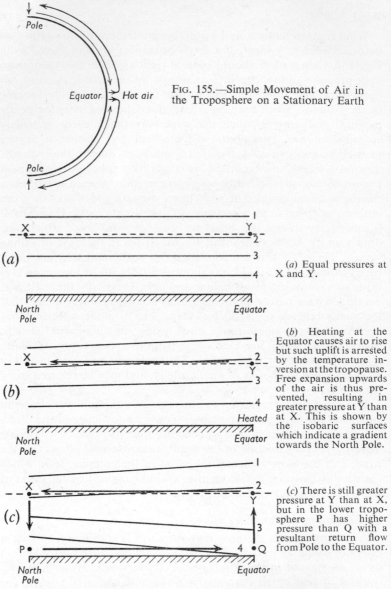

FIG. 155.—Simple Movement of Air in the Troposphere on a Stationary Earth

(a) Equal pressures at X and Y.

(b) Heating at the Equator causes air to rise but such uplift is arrested by the temperature inversion at the tropopause. Free expansion upwards of the air is thus prevented, resulting in greater pressure at Y than at X. This is shown by the isobaric surfaces which indicate a gradient towards the North Pole.

(c) There is still greater pressure at Y than at X, but in the lower troposphere P has higher pressure than Q with a resultant return flow from Pole to the Equator.

FIG. 156.—Simple Circulation on a Stationary Earth as shown by Isobaric Surfaces and Pressure Gradients (in Northern Hemisphere)

1, 2, 3, and 4 indicate increasing values of pressure on isobaric surfaces, – – marks a uniform height above surface of earth.

with no temperature inequalities, there would be stationary layers of air which would exert a uniform pressure over the whole surface (Fig. 156 (*a*)). Let us suppose that heating occurs at one point of this surface, as, for instance, at the Equator, then the isobaric surfaces are disturbed above this point (Fig. 156 (*b*)), so that a pressure gradient is established at relatively high levels and a flow of air is initiated towards each pole as shown by the arrows. However, in time a further disturbance must develop by the collection, at the poles, of air that has been transferred from the equatorial regions, and this will result in an increase of air over the poles. The isobaric surfaces would then appear as shown in Fig. 156 (*c*), with the consequent development of a return flow of air towards the Equator along the surface of the earth. It must be emphasized that such movements are limited in a vertical direction by the tropopause (see p. 197).

Ferrel's Law

This simple system is upset by the fact that the earth is not stationary, but is rotating on its axis, and air moving in any direction in either hemisphere does not follow a consistent track, but suffers deflection. This effect of the rotation of the earth is summarized in Ferrel's Law, which states that any object that moves freely (i.e. under natural conditions) over the surface of the earth will be deflected to the *right* of the path in which it is travelling in the northern hemisphere and to the *left* of its path in the southern hemisphere. It is important to understand clearly what is meant by to the right (or left)

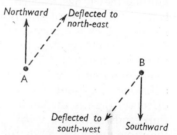

Fig. 157.—Deflection of Winds (in Northern Hemisphere)

of the path, since confusion may arise in applying this law. Fig. 157 shows two objects, A and B, which set off, in the northern hemisphere, to move northwards and southwards respectively. The rotation of the earth causes deflection from these initial tracks as indicated by the broken lines, so that A is eventually moving towards the north-east and B towards the south-west. Both have moved away to the right of the path in which they are travelling, despite the apparent leftward trend of B. It may help the reader to understand this point quite clearly if the full arrow is regarded as a vehicle moving along a road which it leaves, irrespective of obstacles such as hedges, by gradually turning off to the right.

The explanation of this deflection is far from simple. On a rotating globe all points are moving with the same angular velocity, that is, they move through 360° in a day and travel round in a circle. The higher the latitude of the point the smaller is the circumference of its circular path. Thus a point on the Equator moves round, in an easterly direction, at a speed or with a circumferential velocity of 25,000 miles in twenty-four hours, whilst a point at 60° N. or 60° S. moves round a path of only 12,500 miles in the same time. Thus the circumferential velocity is greatest at the Equator, and decreases with increasing latitude to nil at the poles of the earth.

As the earth rotates from west to east, an object which starts to move northwards from the Equator has an initial eastward speed of 25,000 miles per day, but as it proceeds towards the North Pole it acquires a slower eastward speed. The reduction in speed of a mass of air moving in this manner is produced by frictional drag in its lower layers which are in contact with the surface of the earth. When this air has moved from the Equator into a higher latitude, although its eastward speed has been retarded by this frictional drag, it is not reduced to the value of the circumferential velocity of the latitude that it has reached. In other words, it is travelling slightly faster in an eastward direction than a point moving on the circle of latitude which is immediately below it. Thus northward-moving air tends to be deflected eastwards (or to the right). Conversely, south-ward-moving air in the northern hemisphere should gradually gain in circumferential velocity, but frictional drag prevents a complete adjustment, so that it is travelling slightly slower in an eastward direction than is the surface of the earth immediately below it. It thus tends to have a westward deflection (to the right). Obviously the same argument holds for the southern hemisphere, but the deflections are leftward. The above argument does not explain that even when the initial movement of the object is in an eastward or a westward direction, there is still a deflection. However, it can be shown that precisely similar deflections, that is, to the right or the left according to the hemisphere, will occur. It may be noted also that at the Equator there is no deflection at all.

Since these deflections occur, it is necessary to regard the upper arrow in Fig. 156 (c) as representing a wind which is not moving to the north, but more eastwards than northwards, whilst the lower arrow shows a wind moving more to the west than southwards. Remembering that a wind is named from the point of the compass from which it is blowing, there would be winds with westerly components at high levels and winds with easterly components at low levels. However, another factor must be considered, namely, the establishment on dynamical grounds of a belt of high-pressure air at 30°–35° of latitude (the Horse Latitudes) in each hemisphere. The control which sets up these belts of high pressure in the subtropics is that of the rotation of the earth on its axis. Each hemisphere is therefore divided by this high-pressure air into two parts, and air moves both polewards and equatorwards from each of the Horse Latitudes. This means that in high latitudes in each hemisphere air will move polewards and be deflected so as to give a westerly wind, whilst in low latitudes air will move equatorwards and be deflected as an easterly wind. The two parts of each hemisphere which are separated by these subtropical belts of high pressure are very nearly equal in area, which suggests that a balance is thus struck between the area influenced by easterly winds and that which is under the control of the westerly winds. This division does not take account of the polar caps where high-pressure air resides over the surface, and from which air moves out in a westerly direction as a result of the operation of Ferrel's Law.

Pressure Belts and Wind Systems

It is now possible to consider that in the lower layers of an atmosphere which is resting on a rotating globe of uniform surface, there would arise several belts of pressure with their related winds. Fig. 158 (a) shows the globe with its five main pressure zones, to which must be added the circumpolar low-pressure belts. The latter must inevitably arise, since in between the polar caps of high pressure and the subtropical belts of high pressure there must be air at lower pressure in general. Fig. 158 (b) shows the winds which should develop as a result of the pressure gradients that exist between adjacent pressure belts, without taking into account the effects of Ferrel's Law. The consequent deflections, according to Ferrel's Law, are shown by the arrows in Fig. 158 (c). It will be seen that there are six groups of winds, and the name of each set is shown on the left of the diagram.

The simple circulation envisaged in Fig. 156 (c) must now be revised. Instead of a continuous equatorward flow of air at the surface of the earth from each pole, there are three distinct wind belts in each hemisphere. The circulation within the troposphere is probably somewhat as shown in Fig. 159 although it must be understood that no final interpretation of this circulation has been given by meteorologists.

There is another law which links wind directions and pressure distributions, known as Buys Ballot's Law. In Fig. 158 (c) any wind arrow in the northern hemisphere is in such a position that if one

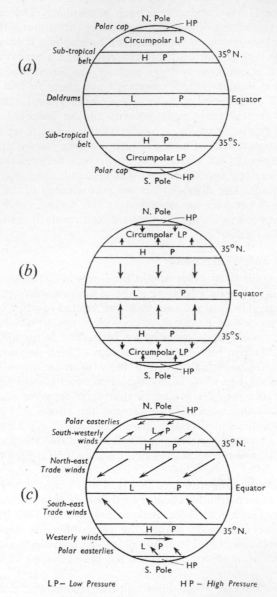

LP— Low Pressure   HP — High Pressure

FIG. 158.—Wind Systems of the World

imagines that one is standing with one's back to the wind which it represents, then higher-pressure air lies to the right and lower-pressure to the left. In the southern hemisphere the converse is true.

The ideal wind system or planetary circulation which is shown in Fig. 158 (c) is far from being realized in actual world conditions, although it is not impossible to recognize, from a study of wind charts of the world, that all of the six winds do prevail over quite large areas of the earth's surface. Departures from this simple system are fairly common, and to give the picture greater reality, we must consider once again the effects of the large continental and oceanic areas. Just as with its pressure distributions so with its winds does the southern world approach more closely to the simple pattern, because of its preponderant ocean surface, which gives that hemisphere greater uniformity than has the northern world. The diagrams in Fig. 160 attempt to show with reasonable accuracy the state of affairs which exists as a result of this uneven distribution of land and water in the two hemispheres. The land masses are represented by an isosceles triangle with its base in the

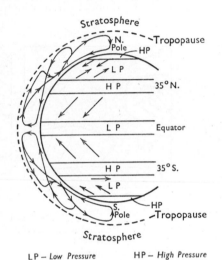

LP – Low Pressure   HP – High Pressure

FIG. 159.—Circulation within the Troposphere

Note that the height of the Troposphere is not shown to scale.

north and its apex opposite to this base projecting into the southern hemisphere. Such a triangle represents the combined American continents or the Old World, that is, Eurasia with Africa, or with Australia and New Zealand.

Several observations must be made on this diagram. Firstly, the pressure distributions are in accord with the world distributions which have been discussed. Secondly, it is necessary to explain briefly the clockwise and anti-clockwise circulations which are indicated by the wind arrows around the several ovals of high and low pressure in different parts of the diagram. If one considers such a cell of high pressure, then pressure gradients will exist in all directions from its centre to its periphery (Fig. 161 (a)), whilst in the case of a low-pressure area there will be gradients in the opposite sense (Fig.

161 (*b*)). Applying Ferrel's Law, the directions of the winds which arise from these gradients will produce what amount to clockwise or anti-clockwise systems, the sense being dependent on two factors, the interior pressure of the particular cell and the hemisphere in which the cell is situated. Fig. 161 (*c*) shows the four cases which arise, although there are, strictly speaking, only two distinct types of circulation, cyclonic around low pressure and anticyclonic around high pressure. Thirdly, in the northern part of the triangular land mass the marked effects of continentality on the pressure values in

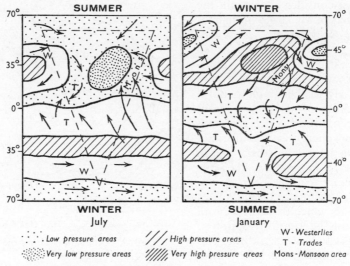

FIG. 160.—Pressure and Winds as modified by Land and Sea
(*After Kendrew*)

the two seasons and the appearance of the monsoon area on its eastern side must be stressed. It must be observed, in connection with the latter feature, that whereas in most parts of Fig. 160 the winds compare quite well with those shown in the diagram of the planetary circulation of Fig. 158 (*c*), in the monsoon area in the summer season there is a complete divergence from the ideal pattern. Whereas in winter the continental high pressure creates gradients which result in north-east winds blowing over the tropical area of the north, in summer the very opposite holds, and instead of north-east winds there are winds from the south, south-east, and east over this eastern part of the land mass. This almost complete reversal of the wind directions constitutes the chief characteristic of the monsoonal climate, which is probably the most outstanding of all the effects of continentality. The Indian climate is unique in this respect, but several other areas of the world are affected in a similar way. It will be noticed, for instance, that in the northern winter season the north-east trade winds transgress the equator, and upon entering the southern hemisphere they are deflected to the left, according to Ferrel's Law. This means that some areas, such as northern Aus-

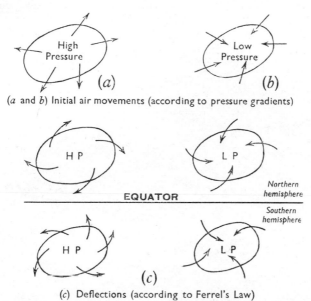

(a and b) Initial air movements (according to pressure gradients)

(c) Deflections (according to Ferrel's Law)

FIG. 161.—Circulations in 'Highs' and 'Lows'

tralia, have north-west winds instead of the normal south-east trade winds.

Before leaving the topic of winds and pressure, a final warning must be given to the reader. Vast areas of the global surface are but slightly known, and the picture which has been built up is based on the average conditions. Hence the reader must not be unduly surprised at finding apparent and real contradictions to some of the facts that have been presented in this chapter. For instance, no account has been given as yet of the cyclonic disturbances which are of such regular occurrence in the circumpolar low-pressure belts in each hemisphere. These disturbances, as will be shown in Chapter XIII, result at times in the establishment of systems of winds which blow, in turn, from almost all points of the compass. Thus the zones which are habitually regarded as being affected by the westerly winds may often have winds other than westerly ones. Even so, winds which have a westerly component (i.e. north-west, west, and south-west winds) are most common. Once again the southern hemisphere is more uniform, and in its circumpolar belt of low pressure westerly directions prevail over all other wind directions. The effect of these prevalent west winds is so pronounced that in the days of sailing-ships the belt of the oceans lying between 40° S. and 50° S. was called the 'Roaring Forties'. Frequently, the 'Roaring Forties' are made synonymous with the 'westerly winds' in the southern hemisphere, but this is an unfortunate use of the term.

CHAPTER XII

WATER VAPOUR IN THE AIR AND CONDENSATION FEATURES

Humidity

Humidity means the water vapour content of the air. Normally, air is capable of receiving evaporated water (or water vapour) from a water surface. The extent to which evaporation goes on depends upon the air temperature, winds promoting evaporation, the amount of water vapour present in the air, and on the supply of water for evaporation at the time. At any given temperature, high or low, there is a limit to the amount of water vapour which a given volume of air can absorb. When this limit has been reached, the air is said to be saturated. At high temperatures much water can be held as water vapour in a given volume of air before the saturated state is reached, whereas at low temperatures only a relatively small quantity is required to saturate the same volume of air. It is possible to determine by physical experiment how much water in the form of vapour is in the air at a given time, and also how much water could be held as vapour if the air was saturated, assuming that there was neither an increase nor a decrease of air temperature. The relation between these two quantities is expressed as a percentage, and is known as the relative humidity of the air.

Suppose a given volume of air at 50° F. contains 40 grammes of water as vapour, and let us suppose that this air at 50° F. could contain (if saturated) 50 grammes of water as vapour. Then the relative humidity of this air would be $\frac{40}{50}$ or 80 per cent. Obviously, if another 10 grammes of water were allowed to evaporate into this air at 50° F., it would become saturated. But this air at 50° F. containing 40 grammes of water as vapour may be saturated in another way. Let us suppose that at 40° F. this same volume of air would be saturated when it held 39 grammes of water as vapour. Then, if this air at 50° F. with 40 grammes of water as vapour were cooled to 40° F., it would become saturated, and it could hold no more than 39 of these 40 grammes of water. Therefore, there would be 1 gramme of water as vapour to spare, and this would return to the liquid state in the form of small drops of water. The values quoted above are hypothetical ones. The amount of water held by a given volume of air, irrespective of saturation, is known as its absolute humidity.

All air contains some water vapour, although in the hot deserts, where high temperatures prevail and where there is little surface water to charge the air with evaporated water, the extent to which the process of saturation has proceeded is often very slight. The relative humidity of desert air is frequently as low as 5 per cent., but it must be borne in mind that with such high temperatures it is possible, if evaporation could go on indefinitely from a water surface, for a very large quantity of water to be assimilated in the form of vapour before saturation occurred in a given volume of air. By the very nature of desert zones, with their paucity of rainfall, and therefore their lack of extensive water surfaces, the chances of the air becoming saturated are remote.

Measurement of Humidity

In modern meteorology the calculation of the water vapour content of the air is of prime importance in the unravelling of some problems. The measurement of relative humidity is made in several ways, but there are only two instruments which need be described.

(a) The Wet and Dry Bulb Thermometer

This instrument consists of two thermometers of the same pattern, but the bulb of one is kept moist by a muslin bag served by distilled water held in a small reservoir; the other thermometer is used normally. Into air which is unsaturated evaporation of water will take place from the muslin around the wet bulb thermometer and, since water requires heat for the purpose of evaporation, this is extracted from the bulb of the thermometer with a resultant depression of its reading. The difference between the readings of the two thermometers, which are usually housed in a Stevenson screen (thus reducing evaporation by both wind and direct sun heat), is an indication of the relative humidity of the air. A small difference indicates a high value of the relative humidity and vice versa. If the air is saturated there will be no difference between the readings of the two thermometers, as no evaporation can take place from the muslin of the wet bulb thermometer. Tables of figures are provided so that the relative humidity may be determined from any pair of readings that are recorded by the two thermometers. Wet bulb readings alone are of value, since when these are regularly over 70° F. in tropical zones white men are liable to the physiological danger of heat stroke.

(b) The Hair Hygrometer

The second instrument for measuring relative humidity makes use of hair that has been made non-elastic by stretching under load.

Such hair will increase its length with an increase of the relative humidity of the air and vice versa. This effect can be used to construct a direct-reading instrument, so that the value of the relative humidity is shown by a pointer which moves over a scale. The instrument is liable to develop mechanical inaccuracies after a period of use, and there is need of fairly frequent recalibration against a wet and dry bulb thermometer.

Dry and Saturated Air

Air which is not saturated may be regarded from a meteorological point of view as dry air, and it is reasonable to say that cold or hot air which is dry is more tolerable and less dangerous to human beings than either cold or hot air which is nearly or completely saturated. Dry air is capable of becoming saturated if its temperature is reduced to what is known as the dew point, when its water vapour content is just sufficient to produce saturation at that lower temperature. Further cooling results in the return of some of the vapour to the liquid state. The processes by which such cooling is effected naturally are of importance, since they can lead to the formation of clouds, mist, and fog, as well as to the production of rainfall and other forms of precipitation. If a mass of dry air is forced to rise from the surface of the earth to higher levels of the troposphere (or lower atmosphere) then a series of changes occurs:

(i) The air will move to a level where the pressure of the atmosphere is lower, since there is less weight of air upon it at these higher levels.

(ii) This reduction in pressure causes the rising air to expand.

(iii) Since air is a gas it will cool as a result of such expansion. This cooling is known as an adiabatic change because it is internal to the ascending and expanding air, and not because the air is moving into colder regions of the troposphere.

Adiabatic Lapse Rate

In discussing the temperature of the air, it will be recalled that there is an average lapse rate in the troposphere of 1° F. for every 300 feet rise. Dry air, in the sense that it is not saturated, cools more readily than this as it rises through the troposphere and expands. It has an adiabatic lapse rate of 1° F. for a rise of 187 feet. This difference in the lapse rates may be shown diagrammatically as in Fig. 162. At the level XY the temperature of the troposphere has fallen from T, at ground level, to T', whilst the rising air has cooled from T to T", a lower temperature than T'.

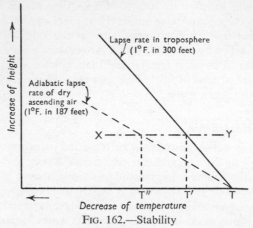

FIG. 162.—Stability

In Figs. 162, 163, and 166–72 the various curves are not drawn to the same scale. They are intended to show some possible changes in the behaviour of ascending air under various controls. Where the letters T, T′, T″, etc., are used, it does not necessarily follow that they represent the same temperature in different diagrams. Each diagram must be regarded separately in this respect.

Stability

Whereas the adiabatic lapse rate for dry ascending air may be assumed constant at all times, the lapse rate in the troposphere may vary. It may be greater or less than the average rate of 1° F. in 300 feet, or, in other words, a fall of temperature of 1° F. will occur with a rise of less than 300 feet or more than 300 feet, respectively. Let us examine air which is rising through a troposphere that has an average lapse rate (i.e. of 1° F. in 300 feet). Returning to Fig. 162, let us assume

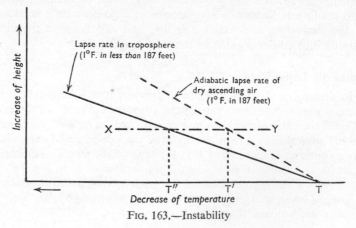

FIG. 163.—Instability

that some air has risen through the troposphere from ground level to the level of XY. Now, at all stages in its ascent it can be seen that the temperature of the rising air is lower than that of the surrounding troposphere. As soon as the air leaves the ground, where it has the same temperature as the base of the troposphere, this difference of temperature appears. Now, the rising air not only becomes cooler, but also it becomes denser than the surrounding air of the troposphere as it rises, and this rising air would tend to sink, so that it returns to ground level. In fact, unless some external force is at work, the air cannot rise because of this difference in temperature and density. Under these conditions there is a state of stability as far as rising air is concerned.

Instability

Now let us see what happens if the lapse rate in the troposphere exceeds the constant adiabatic lapse rate of dry air (i.e. the lapse rate in the troposphere is 1° F. in *less* than 187 feet), admittedly a very unusual state of affairs. Fig. 163 shows such a condition, and once

Official United States Navy photo

FIG. 164.—Cumulus Cloud

This cloud has been produced by convection currents caused by a fire in a cane sugar plantation at Pearl Harbour, Hawaiian Islands.

again we will examine the respective temperatures in the troposphere and rising air at ground level and at the level XY. The temperature in the troposphere has fallen from T to T″, whilst the ascending air has cooled by expansion only from T to T′. In this case at all stages in its ascent the rising air is warmer and, of course, less dense than the surrounding air in the troposphere. Consequently its ascent will continue, since the denser surrounding air tends to force it up to greater heights. This rising unsaturated air is said to be unstable and, under the circumstances described, it would appear that it could rise indefinitely owing to its instability in the troposphere.

Convection

One common cause for the ascent of air is an initial rise in its temperature at ground level through local heating, which thereby induces convection (Fig. 164). Some of the dry air is then forced up-

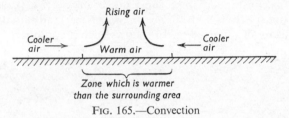

FIG. 165.—Convection

wards by the surrounding air, which is slightly cooler and denser, pressing on it (Fig. 165). Let us now follow the course of air which has been set moving upwards by convection through a troposphere which has an average lapse rate (of 1° F. in 300 feet). Fig. 166 demonstrates its history. Clearly up to the level PQ the rising air would be warmer than the surrounding troposphere, whilst above this level it

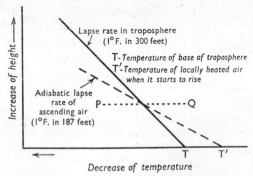

FIG. 166.—Unstable Air rising to a Level where it becomes Stable

would be cooler. Consequently the air would rise to the level PQ, but if it went higher it would sink back to this level. In other words, conditions are unstable below PQ, but above they are stable as regards the rising air. Thus it would seem that under conditions of instability rising air will ascend until it cools by expansion to the temperature of the surrounding troposphere when its ascent is arrested.

Changes in Ascending Air

However, it is possible for ascending air to rise above this level, PQ in Fig. 166, if its rate of cooling by expansion is retarded. Before reaching this level the rising air may have cooled sufficiently for the condensation of water vapour to take place within it. The water so formed collects in droplets around particles of dust and other

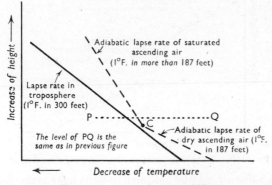

FIG. 167.—Increasing Instability

materials which are available, and a cloud is formed. Two other changes occur which are of importance. Firstly, when the water vapour condenses to water heat is liberated in a comparable amount to that which was required to evaporate the same amount of water. This liberated heat is helpful, since it reinforces the upward movement of the air in much the same way that convection set it off from the ground. Secondly, the heat so released reduces the rate of cooling of the air with its further ascent as compared with the adiabatic lapse rate for dry (unsaturated) air. The saturated or wet adiabatic lapse rate after condensation occurs is not constant, but is usually about 1° F. in 370 feet. The result of this change is shown in Fig. 167. Air rises to C, where condensation takes place, and continues to rise indefinitely above this point. Obviously unstable conditions allow it to rise to C, and its decreased lapse rate above that point makes the conditions relatively more unstable, because there is an increasing difference of both temperature and density between the rising air and the surrounding troposphere, so that the air does not halt at the level of PQ as in the former example (see Fig. 166).

It would appear that the rising air is so unstable above C in Fig. 167 that it would go on ascending, since the surrounding troposphere becomes progressively colder than it. Such is not the case, because higher up the water droplets in the rising air freeze, and again the adiabatic lapse rate changes, when it approximates to that of dry air. Fig. 168 shows the effect of this last change on the cooling of the ascending air. From C to F the air cools at the reduced rate for saturated or wet air. At F freezing occurs and further cooling is shown by the slope

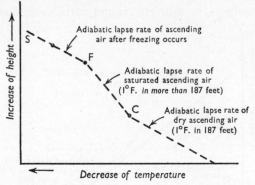

FIG. 168.—Instability reduced by Freezing Conditions

FS,[1] and at S this air may be assumed to be as cool as the surrounding troposphere. It would be under stable conditions if it rose above S, so that it is actually halted at this point. In passing it should be noted that the freezing of water in the rising air does not necessarily occur when its temperature has fallen to 32° F., since water may remain in the liquid state at temperatures well below normal freezing-point. This super-cooled water will be referred to in relation to the formation of heavy rainfall (pp. 202-203).

Before leaving this account of air which rises because of instability, some further possibilities must be investigated. In all the examples which have been illustrated the lapse rate in the troposphere has remained constant, but that is unusual. It frequently changes, and at higher levels in the troposphere the lapse rate almost invariably decreases. In Fig. 169 air is shown rising through a troposphere whose

[1] Actually, a release of heat occurs with the freezing of the water droplets and temporarily the adiabatic lapse rate would decrease above F. The slope would become practically vertical for a very short distance on the graph.

lapse rate decreases above Z, and the rising air would be halted at X and not higher, as it would if the lapse rate in the troposphere had remained unaltered. Sometimes the lapse rate in the troposphere

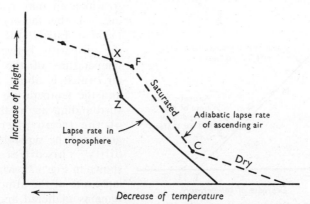

FIG. 169.—Ascent of Unstable Air halted by a Decrease in the Lapse Rate in the Troposphere at a High Level

changes so that there is actually an increase of temperature above a certain level, known as a temperature inversion. The effect of an inversion is shown by Fig. 170, where it will be seen that stable conditions are imposed upon the rising air at an even lower level than

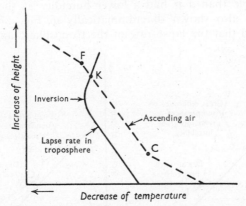

Fig. 170.—Ascent of Unstable Air halted by an Inversion

before. In this particular example unstable air rises to C where condensation occurs, and it proceeds with a reduced adiabatic lapse rate (for saturated air) but does not reach heights where freezing occurs because of the halt which is forced upon it at K, above which stable conditions prevail.

The various examples which have been discussed are merely a few of the many possible histories of ascending air. In addition to the variations of the lapse rate in the troposphere, which affect the limits to which air may rise, there are two other factors to bear in mind. Firstly, the greater the disparity between the temperature of the heated air, which is about to rise, and the temperature of the surrounding air at the base of the troposphere, the greater is the possible instability. This difference is shown in Fig. 171, where the lapse rate of the atmosphere remains unaltered, and whilst warm air at T_I will rise to L, the warmer air at T_{II} will rise higher to M before it becomes stable. Secondly, the higher the humidity of the air that is about to ascend, the lower will be the condensation level beyond which it proceeds at the reduced adiabatic lapse rate for saturated air, and this means that it is relatively more unstable than if it had a lower humidity at the start. This difference is also shown diagrammatically in Fig. 172, in which it is assumed that the lapse rate of the troposphere and the initial

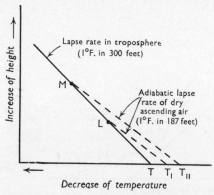

FIG. 171.—Varying Instability, dependent on the Original Temperature of the Rising Air

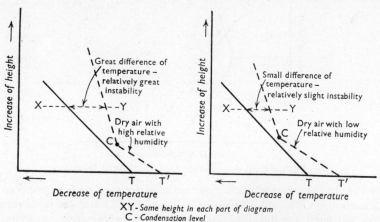

FIG. 172.—Varying Instability, dependent on the Humidity of the Rising Air
(In these diagrams 'dry air' means that it is unsaturated)

temperature of the rising air are constant. These various factors, the lapse rate in the troposphere, the initial temperature of the rising air, and its humidity, are all variable, and since they interact with one another it is obvious that the possibilities are limitless.

Tropopause

It has been mentioned that the lapse rate in the troposphere usually decreases at higher altitudes. Above a certain level, which is not constant in height, the temperature of the atmosphere becomes steady or it may increase to some extent. The tropopause, as this particular inversion level is called, is the effective limit of ascending air, assuming that unstable conditions had persisted so that, under favourable circumstances, air could rise to these heights. The level of the tropopause is higher over low latitudes than it is over high latitudes of the earth's surface, and it varies with the season of the year. On the average its height above the surface of the earth is about ten miles at the Equator and less than seven miles nearer the poles. The tropopause divides the atmosphere into a lower layer, the troposphere, and a higher layer, the stratosphere. In the former cloud formation is possible because of unstable conditions which permit upward movement of air, whilst in the latter clouds are absent, thus making this layer attractive for aeroplane flights. In addition, the stratosphere is free from fluctuating pressure distributions which cause air pockets in the troposphere.

Condensation

The condensation of water vapour which occurs when rising air reaches a level where its temperature produces saturation is not effected under normal atmospheric conditions with quite the simplicity which has been suggested. In fact, the formation of water drops from the water vapour is apparently a rather complex process in a troposphere which is relatively pure and clean. Satisfactory conditions for condensation are determined largely by the number and the type of solid particles which are in suspension in the air. It would appear that progressive condensation is assisted by the presence of dust, smoke particles, grit and minute salt crystals, also by very small electric charges on the droplets of water. These several agents clearly encourage the formation of large drops of water by the steady growth of droplets. Some authorities regard the small salt crystals as the most valuable of all, since it has been shown by analysis that they are present in all raindrops, whether they fall in oceanic or continental locations. The particles of solid material which are found in industrial smoke are particularly hygroscopic (that is, they will absorb water from vapour very readily) and are capable of inducing condensation when the air has a relative humidity of only 80 per cent. and not 100 per cent., which is the normal value for saturated air. Condensation is manifest in the atmosphere as fog, cloud, and by the precipitation of rain, hail, snow or sleet, dew, and hoar frost.

Clouds

Clouds are divisible into two main groups which are related to the stability and instability of the troposphere:

1. Layered clouds show a fairly horizontal stratification and little

Aerofilms Ltd.

FIG. 173.—Heap and Layered Clouds

Cumulus clouds in the foreground and stratus clouds in the background.

vertical extension. The most common form of layered cloud is known as stratus (Fig. 173). Fog is closely related to this form of cloud, and if fog is forced to rise from the surface of the earth it will develop into low stratus cloud. A stable troposphere is essential for the development of layered clouds. The stratification and the slight depth of the cloud are due to the fact that rising air is made to halt at the condensation level by the stability of the troposphere. Not infrequently a line of layered clouds marks the level of a temperature inversion. If dry air rises to such an inversion level without condensation of water, it is usual to see a continuous line of smoke haze in place of stratus cloud.

2. Heap clouds are developed when there is considerable uplift of air by convectional currents assisted by instability which causes these clouds to have a much greater vertical extension than the layered forms. There is a distinction amongst heap clouds which is dependent upon whether the ascent of the air is limited or unlimited. In the first case the vertical development of the cloud is limited by the rising air entering a stable layer of the troposphere, usually at an inversion level. On the other hand, heap clouds may form in a troposphere in which instability persists to great heights, and then the clouds may show an upward extension which places their tops as high as the level of the tropopause. One of the most common heap clouds is known as cumulus (Fig. 164). Heap clouds, with their marked vertical extension, are not individually stratified, but groups of these clouds may show a certain amount of layering (Fig. 173).

There are many cloud types, but the majority show either cumuliform features or stratiform features. Clouds may be grouped, according to their level in the troposphere, into high, medium, and low clouds. The cloud diagram (Fig. 174) shows the appearance of the most common clouds and their average levels above the earth's surface. The inner part of the diagram may be regarded as a triangle having cirrus, stratus, and cumulus clouds at its apices, and intermediate types along its three sides. Cirrus clouds are very high and resemble feathers or mere wisps of cloud which are composed of ice crystals and not of water drops (Fig. 175). Some of these high clouds may be stratiform (cirro-stratus), in the form of a thin, whitish sheet, and others may be cumuliform (cirro-cumulus), composed of many small rounded cloud masses which are arranged in lines, so as to resemble the sand ripples on a seashore. Strato-cumulus cloud is an intermediate form to stratus and cumulus (Fig. 233). Like cirro-cumulus cloud, it consists of detached masses of cumuliform cloud which are arranged very often in lines, although the individual masses often appear to overlap one another. Lying between the very high cirrus group and the low clouds are the alto-cumulus and the alto-stratus types, which are similar to, but thicker than, cirro-cumulus and cirro-stratus clouds respectively. Two other clouds appear on the diagram and both may have very great vertical development, in that they have bases which are low in elevation and yet their tops may be

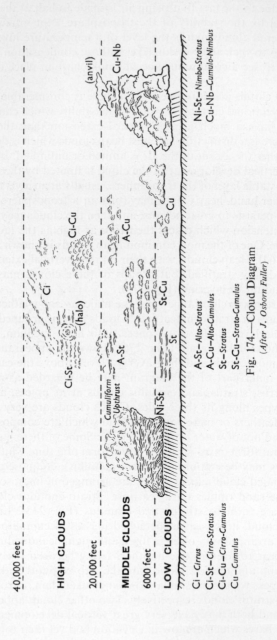

40,000 feet

HIGH CLOUDS

20,000 feet

MIDDLE CLOUDS

6000 feet

LOW CLOUDS

Ci – Cirrus
Ci-St – Cirro-Stratus
Ci-Cu – Cirro-Cumulus
Cu – Cumulus

A-St – Alto-Stratus
A-Cu – Alto-Cumulus
St – Stratus
St-Cu – Strato-Cumulus

Ni-St – Nimbo-Stratus
Cu-Nb – Cumulo-Nimbus

Fig. 174.—Cloud Diagram
(After J. Osborn Fuller)

as lofty as some of the high clouds. These clouds are called nimbo-stratus and cumulo-nimbus. The former has a fairly low stratified base, with a shapeless and massive top, whilst the latter appears from below rather like a nimbo-stratus cloud, but it towers up to very great heights with cumuliform characteristics and often with a peculiar flat top, elongated at one end like an anvil, which marks an inversion. Both of these clouds are productive of heavy rainfall, and the cumulo-nimbus cloud is associated with thunderstorms (Fig. 198).

The type, height, and thickness of clouds depend on various factors.

Clark Collection, Royal Meteorological Society

FIG. 175.—Cirrus Cloud

Nearly saturated air rising from the ground will tend to produce clouds at low levels, whereas air of low relative humidity may rise very high before there is condensation. Warm air which can hold much water vapour is therefore capable of producing very dense clouds compared with air which is initially at a low temperature. Fairly warm and dry air may have to rise to very great heights before only thin stratiform cloud is formed. Warm and humid air may remain cloudless at considerable heights until it glides over a layer of colder air which chills it sufficiently to produce clouds at its base. Small cumulus clouds at low level are usually indicative of fair weather, and are frequently associated with strong surface heating of the land below, particularly in summer, when such clouds show a maximum development about mid-afternoon

It is probable that strato-cumulus, alto-cumulus, and cirro-cumulus clouds are formed by the disintegration of earlier-formed stratus, alto-stratus, and cirro-stratus clouds respectively. The stable conditions are destroyed by convection, and the clouds are broken up into a series of cloudlets. Cirrus clouds are very high and thin, because convective conditions have reached the highest limits of the troposphere where the slight condensation which is possible is in the form of ice particles rather than water drops. Ice crystals are responsible for the haloes produced by the light of the sun or moon shining through various types of thin high cloud.

Stratus cloud may develop into nimbo-stratus by a considerable vertical extension, and this results in quite heavy rainfall from the upper layers of the cloud, although the rain appears to come from the base of a layered cloud. In cumulo-nimbus cloud the cumulus type has developed on similar lines, but to even greater heights, where its top is flattened or stratified because of the influence of a very elevated temperature inversion. It is quite common for great upthrusts of cumuliform cloud to rise out of a cloud mass. Such upthrusts indicate the presence of a local column of air which is unstable. These upthrusts are much better seen from aeroplanes than from ground level. Although rain may fall from most clouds it is only from those of great vertical development, such as nimbo-stratus and cumulo-nimbus, that heavy rain can be expected. From stratus clouds, in general, and from many cumulus cloudlets there is, at best, light rain or just drizzle.

Precipitation

Heavy and even moderately heavy rain is dependent upon the condensation of water vapour primarily as ice rather than as water, and usually air must be lifted to very great heights in order that its temperature may fall below freezing-point. Water may remain unfrozen at temperatures well below freezing-point, and from a mass of air which has risen high enough it is usual for ice crystals to form as well as water droplets when condensation takes place. It is also possible for very cold water to evaporate and condense on these ice crystals as ice instead of as water. This process leads to a considerable growth in the size of the ice crystals, which gain in weight to such an extent that they fall through uprising air and melt at lower levels to form large drops of water, and as these pass still lower, they tend to pick up more water from cloud droplets by collision, so that their size is increased still more although there is a limit to this growth (see p. 228).

These are regarded as the essential conditions for the production of heavy rainfall, and it is only in clouds with a great vertical development that these conditions can arise. From lower clouds precipitation will be restricted to light rain or drizzle, since the elevation is not great enough for the formation of ice with the consequent formation of large drops of rain.

It is of interest to note that rain has been produced by the introduction of frozen carbon-dioxide crystals into the upper parts of a cloud from which no rain has fallen. In Australia this treatment was successful. Shortly after the insertion of this 'dry ice' from an aeroplane into a cloud top rain was seen to fall from the cloud. In England, in 1947, the same experiment proved unsuccessful in an attempt to end a period of prolonged drought. Although scientists have unravelled some of the problems of rainfall production, they have not found out all the conditions which will precipitate rain from clouds by artificial means.

Water may remain liquid at temperatures below 32° F., when it is referred to as super-cooled water. Under — 4° F. it usually freezes to ice, but water vapour condenses directly as snow in such cold air. Much rain actually starts to fall as snow or hail, and then melts to water in passing through warmer layers of the troposphere. When snow falls at ground level it has, in many cases, started as rain from a high cloud, and has changed to snow on the way down. This statement, which may be surprising, will be clarified when the reader has perused the next chapter, but for the present it must suffice to say that the precipitation of snow occurs when a very much colder layer of the troposphere, lying near to ground level, is overlain by quite warm air in the higher levels of which clouds accumulate to give abundant precipitation. Sleet represents a transitional stage between water and snow, or it may be partially melted snow.

The measurement of precipitation and its distribution in the world are discussed in Chapter XIV.

Mist and Fog

Various cooling processes other than adiabatic ones may lead to the condensation of water in the lower levels of the troposphere with the formation of mist or fog. There is no essential difference between mist and fog other than the greater reduction of visibility which occurs with the latter. Evaporation may take place from a water surface, or a shallow surface layer of very humid air which is warm, into a superior layer of cooler air in the lower part of which, after saturation occurs, there will be condensation if water continues to evaporate from below. In this process a very thin mist may form such as gives 'steaming' over a lake surface that is overlain by cold morning or night air. The vertical extent of such mists is slight and they rarely persist for long. At meteorological stations fog is recorded when the visibility is less than 1,000 metres.

Condensation frequently occurs when fairly humid air lies over, or passes over, a surface which is very much colder. Such surfaces are usually cooled through the radiation of heat, which is taking place at all times from the surface of the land and, also, from particles of solid material in the troposphere. During most days insolation more than makes good this loss of heat, but at night, when insolation is absent, much heat is lost from the surface of the land, particularly when the night is cloudless. It is with radiation chilling that the temperature of the air in contact with these surfaces is reduced very readily to the dew point. The water droplets that form may collect on the ground as dew, or, if the temperature is low enough, a cover of hoar frost will be formed. Rime, which is very much coarser in texture than hoar frost, is produced by super-cooled water freezing on a cold surface. In the lowest layer of the troposphere which lies above a chilled surface condensation takes place on solid particles, and fog or mist is formed, at first just above ground level. Such fog or mist may extend upwards if there are slight winds which can produce a certain amount of turbulence that creates a vertical spread of the chilled air. This results in an increase of condensation and a thickening of the fog layer. Obviously dead calm will not give this necessary turbulence, but if the winds are moderately strong, with speeds of more than 8 or 10 m.p.h., their turbulent effect is too great, and the chilled air is dispersed too widely and too rapidly for the temperature of the higher layers to be lowered to their dew point. The density of the fog may be increased by the loss of heat from the particles upon which condensation occurs, since they, too, are capable of radiating heat.

Although such radiation fogs may develop at any season in Britain, they are most common in autumn and winter. In the former season, which is characterized by fairly calm and clear weather of anticyclonic origin (see p. 219), there is an intensification of radiation at night which produces relatively greater chilling, whilst in winter the longer and colder nights are conducive to well-developed fogs. The persistence of a fog is dependent upon the difficulties of its dispersion. The most effective agent for this dispersion is a strong wind, but insolation in the daytime is helpful in creating convection currents which in turn produce vigorous turbulence. In winter, with deeper fogs and less insolation, the period required for dispersion may be prolonged until the warming of the lower layers of the fog is sufficiently advanced. Again, in winter, low temperature means that cold air, which is more readily saturated than warm air, will require greater heating before it can accommodate the evaporated water from a fog.

Fogs in densely populated areas, particularly those which are industrial, are often more pronounced than those that form in rural areas. This is largely explained by the presence of abundant hygroscopic particles in the air of towns that readily cause condensation of water drops from air which is far from being saturated. Observations show that the upper level of many fogs is coincident with a temperature inversion. Such an inversion may be detected on a clear day by a fairly horizontal layer of smoke. An inversion is also responsible for a high fog, which may consist very largely of a layer of densely clustered particles, with or without condensation features. At street level, in a large town, with such a fog visibility is perfect, although the sky above is completely obscured, so that it appears like night. Much of the sun's heat that is required at ground level for fog dispersion is either reflected from the upper surface of the fog layer or absorbed by the particles in the fog, in which case the temperature of the air at the inversion level is increased, thus strengthening the inversion and helping to prolong the fog.

Some fogs are produced by advection (horizontal motion of air in contrast to vertical convection) when warm humid air moves over a cold surface or cold air moves over a warm surface; in either case condensation will take place. Such fogs are common in our winter, when a prolonged spell of radiational chilling and severe frost is followed by mild weather, with winds of oceanic origin. They are rarely persistent over land areas, since the warm air stream will soon cause an increase of temperature in the colder land surface, with a resultant evaporation of the fog. Most sea fogs are advective, and are usually developed in the neighbourhood of cold sea currents, as along the coasts of Peru and California, although they are very common in offshore tropical air masses which have dew points exceeding the sea temperatures. In polar regions pack-ice surfaces produce in much the same way a shallow layer of mist or fog which is called 'sea smoke'.

The fogs off Newfoundland are of particular interest since they form despite the presence of strong winds which normally prevent the formation of fog. In this region the mixing of warm and almost saturated air, blowing from the south, with cold humid air from the north results in very heavy condensation of water vapour as fog. These fogs are certainly related to the temperature contrast in the air over the warm Gulf Stream drift and the cold Labrador current.

CHAPTER XIII

AIR MASSES, DEPRESSIONS, AND ASSOCIATED SYSTEMS

Air Masses

Weather conditions in many parts of the world are dependent upon the interaction of air masses. An air mass is a very large part of the lower troposphere which covers a very extensive area of an ocean or continent known as the source region, and which has certain meteorological characteristics. Important changes may occur in these characteristics as a result of the movement of the air mass from its source region. The usual sources of air masses are the main centres of dispersal of air, and these are essentially the chief zones of high-pressure air. It has been shown that the location of the fairly permanent high-pressure areas is over the oceans in subtropical latitudes and in the polar regions, to which must be added those which develop in winter over the continents of North America and Eurasia (Fig. 178). Air masses may be classed as polar or tropical, according to the temperature at the source, and as maritime or continental, according to the humidity at the source. Polar air masses will be colder than tropical ones, whilst maritime masses will be more humid than continental ones. There may be distinguished four primary types of air mass, namely, polar maritime, polar continental, tropical maritime, and tropical continental, the latter being located almost exclusively in the Saharan region, since this is the one area of continental surface within the tropics which is extensive enough to develop a distinct air mass, although Mexico may be regarded as another source of this type.

When air masses move from their sources, the effects are varied, and the precise history of the changes which occur is not known in the majority of cases with anything like accuracy. In general, a tropical air mass which moves polewards will tend to be cooled by the surfaces over which it passes, and conversely a polar air mass, moving equatorwards, will be warmed. It is, however, only the lower layer of an air mass which is so affected, the upper part remains relatively unaltered in regard to its temperature. These changes, which occur in the vertical distribution of the temperature within air masses as they move from their sources, tend to produce a fairly high degree of stability at the base of a tropical air mass and of instability in a polar air mass.

Fig. 176 shows how the lapse rates within these two types of air mass change with their movement. The full lines represent the lapse rate at the source, and the broken lines show the changed lapse rate after

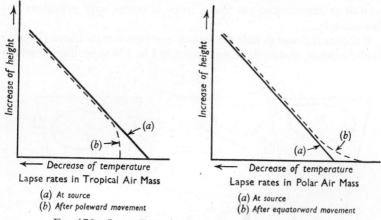

Lapse rates in Tropical Air Mass

(a) At source
(b) After poleward movement

Lapse rates in Polar Air Mass

(a) At source
(b) After equatorward movement

FIG. 176.—Lapse Rates in Tropical and Polar Air Masses

movement of the air mass in question. Reference to Fig. 177 will show the relative instability of the polar air mass, since ascending air would halt at X in the tropical mass, but would rise to Y in the polar

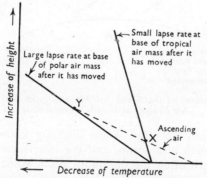

FIG. 177.—Instability at Base of Air Masses
Note scale is greater than that of Fig. 176.

mass. Air masses which move from ocean surfaces on to continental areas, or vice versa, will also be transformed by the changes in their humidity. For instance, when a tropical maritime air mass moves from an ocean to a fairly cool land surface it will tend to become very stable in its lower layers, whilst a polar continental air mass, moving out over an ocean, will rapidly acquire great instability. On the contrary, a tropical maritime air mass moving on to a very hot continental surface will become very unstable with the production of copious rainfall as experienced on many tropical coasts with prevalent onshore winds.

It is over the oceans that the interaction between air masses is most likely to occur, and so there are shown in Fig. 178 three lines at which

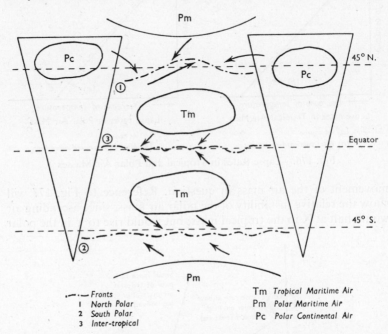

FIG. 178.—Distribution of the Chief Air Mass Sources and the Interaction of Air Masses at Fronts (generalized)

moving air masses converge. These lines are known as the polar fronts of the north and south, and the intertropical front. These various fronts are often discontinuous on a world map, but they usually run in something like an east–west direction, although locally they may trend in any direction. When two air masses, one of tropical and the other of polar origin, meet at the polar front, apart from the changes in each air mass which have been outlined above, there is usually a considerable difference in their respective temperatures and humidities. When the air masses meet, this contrast results in the development of kinetic energy, which, in combination with the effects of the earth's rotation, leads to the formation of a cyclonic disturbance or depression. Into equatorial regions air masses move from north and south, and converge at the intertropical front, but as the opposed air masses are alike in regard to both temperature and humidity because of the similarity of their sources, the subsequent formation of cyclones is practically impossible.

Formation of Depressions in the Northern Hemisphere[1]

The effects which result from the interaction of two air masses of polar and tropical origin respectively are considerable. A subtropical high-pressure zone disperses tropical air in a poleward direction, and in time this must meet air of polar origin which has moved equatorwards. It is usual for such air masses to move in such a way that where

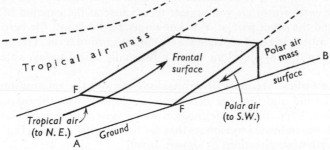

FIG. 179.—A Front and a Frontal Surface (Section of Troposphere)

they meet there are two flows of air which are more or less opposed to one another on either side of a line that may be traced on the surface of the earth (Fig. 179); this line, FF, is called a front. It would appear that contrasted air lies on either side of this front but the line FF is merely the boundary between the air masses at the surface of the earth. A vertical section along the line AB will show that the front FF is the base of an inclined surface which rises from the earth's surface and separates tropical air above from polar air below. The gradient of this frontal surface is about 1 in 100. Although the tropical air has cooled since leaving its source, it is still relatively warm, and the colder and denser polar air forces the warmer and lighter tropical air to rise up the frontal surface. The simple front, FF of Fig. 179, is rarely maintained and is usually modified by the development of a bulge (Fig. 180) which alters the relative direction of the air flows on either side of the front. This bulge upsets locally the

[1] Whilst the following five sections of this chapter describe depressions in the northern hemisphere, the facts given are fundamentally true for depressions in the southern hemisphere.

balance between the two air masses at the surface. It is probably initiated by several causes such as a local increase in the contrast between the temperatures of each air mass whilst, in some instances, relief features on land are responsible. The bulge can be looked upon as a forward thrust of warm air at X, and this is followed by a forward thrust of cold air at Y (Fig. 180).

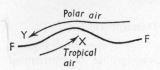

FIG. 180.—The Development of a Frontal Bulge (Plan)

These two thrusts produce a wave, resulting, it is thought, from shear at the frontal surface, and this travels along the front at the surface level of the earth, and in most cases in an easterly direction, irrespective of the hemisphere in which it occurs. With this movement the part of the front at X is known as the warm front, since at the surface of the earth warm air will appear to advance at the expense of cold air, and the other part at Y is known as a cold front, for here cold air is advancing and replacing warm air at ground level. The moving wave constitutes a cyclone or, as it is usually called today, a depression.

The earliest accounts of depressions showed the relationship between wind and pressure distributions, together with the associated weather. Very simply a depression was a system of low-pressure air which was indicated by isobars that formed an oval pattern (Fig. 181), and with wind directions as shown. It will be noticed that Fig. 180 shows a comparable circulation of air although less well developed. Clearly the injection of warmer air at the surface inside the bulge at the front will induce low-pressure air, since it will consist of relatively light air. The later history of the depression depends on the moving wave, which may migrate along the front as a minor disturbance, with little or no change from

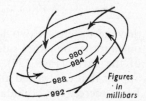

FIG. 181.—Cyclonic Pattern of Isobars (Northern Hemisphere)

its initial state, or it may develop very considerably. If the bulge shows a wave-length of very much less than about 600 miles, it is not likely to develop, but migrates along the front until it is destroyed by frictional effects. When the wave-length is about 600 miles the bulge is likely to evolve into a vigorous depression. The bulge of Fig. 180 becomes much deeper, as in Fig. 182 (a), with further complexity of air-flow directions. The tropical air at the surface of the earth occupies a zone known as the warm sector, which is bounded by the warm and cold fronts. Sections through Fig. 182 (a)

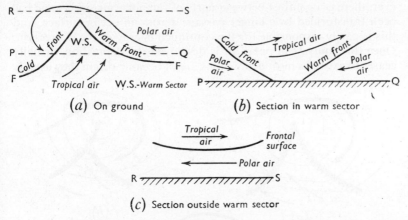

(a) On ground   (b) Section in warm sector

(c) Section outside warm sector

FIG. 182.—Further Development of the Frontal Bulge

will clarify the directions of the air movements in relation to the two fronts (Fig. 182 (b) and (c)). In both sections it will be seen that the warm air is being lifted above colder air.

Occlusions

A further stage in the development of the depression concerns the relative positions of the fronts to one another. As the disturbance progresses eastwards, each front will move in that direction, but the cold front will often overtake the warm front (Fig. 183 (a–c)). The result is that the air of the warm sector is being lifted more and more from the ground. This process is called an occlusion, and it culminates with the elimination of the warm front at the ground where warm air has been completely replaced by colder air. Occlusions are of two types, warm front and cold front. In the former the warm air is lifted up the surface of the warm front, and in the latter it moves up the cold front (Fig. 183 (d and e)).

The low pressure which forms at the centre of a depression in its earliest stages is focused at the junction of the two fronts (Fig. 183). With the evolution of the disturbance and the extension of the occlusion, the focus of low pressure usually remains at the end of the occluded front, whilst the junction of the active fronts shifts from the centre of low pressure; at the same time the low pressure deepens and becomes more extensive (Fig. 183 (c)). It will be noticed (Fig. 182) that the lower layers of air in the depression are mainly of polar origin, and the tropical air at the surface decreases with the shrinking warm sector in the later stage of the history of the disturbance. However, there is a contrast between parts of the polar air since some has been transformed by a longer passage across an ocean surface, and this is partly responsible for the continued energy of the depression. Once this contrast in polar air is destroyed the depression virtually ceases to exist. As most depressions are of oceanic origin they usually

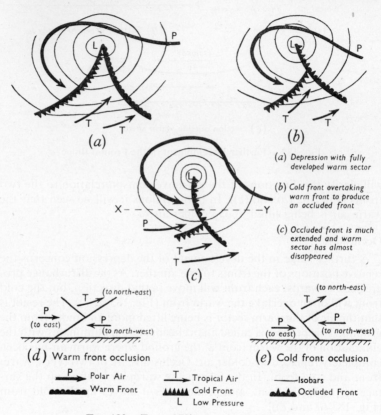

(a) Depression with fully developed warm sector

(b) Cold front overtaking warm front to produce an occluded front

(c) Occluded front is much extended and warm sector has almost disappeared

(d) Warm front occlusion

(e) Cold front occlusion

P→ Polar Air	T→ Tropical Air	—— Isobars
▬▬▬ Warm Front	▲▲▲▲▲ Cold Front	▲▬▲ Occluded Front
	L Low Pressure	

FIG. 183.—Frontal History of a Depression

though not necessarily, die out near to coastal areas or over land masses where the contrast between air masses is reduced to a minimum. The weather conditions associated with the fronts of a vigorous depression are significant to the geographer. Two considerations must be borne in mind when discussing these fronts: (a) the polar air is becoming unstable whilst the tropical air increases its stability, and (b) tropical air is being lifted above polar air in the depression.

Weather at a Warm Front

The warm front shows a remarkable succession of condensation features, either as clouds or in the form of precipitation (Fig. 184). In reading this diagram one must realize that X is a point on the surface of the earth well ahead of the warm front (WF) and that in the course of time conditions as shown at Y in the warm sector will have progressed to X. At the time indicated by the section, tropical air lies high above X, perhaps near the top of the troposphere, where it has

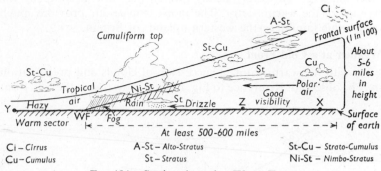

| Ci – Cirrus | A-St – Alto-Stratus | St-Cu – Strato-Cumulus |
| Cu – Cumulus | St – Stratus | Ni-St – Nimbo-Stratus |

Fig. 184.—Section through a Warm Front

been lifted by the wedge of colder air. The stable nature of tropical air in the warm sector is reflected by the strato-cumulus clouds, but the effect of ascent on this air has initiated condensation which has led to the release of heat, convective upthrusts, and the formation of high cloud from which rain is falling. Further ascent of the tropical air up the frontal surface has resulted in the production of less heavy cloud forms which show a succession of strato-cumulus and alto-stratus clouds, passing out to filamented cirrus more or less immediately above X. As the warm front moves towards X it is possible to observe the change in the weather conditions. Ground winds may be easterly or south-easterly, with fairly low temperatures, and there is clear sky with high cirrus cloud in the front of the depression as at X. Some hours later high stratiform clouds are seen advancing from the west, whilst winds are still predominantly easterly in direction at ground level. At lower levels stratus clouds may be seen moving in the polar air stream towards the west. This is in the zone immediately above Z. Later still the clouds become generally thicker and lower and, as the warm front (WF) approaches the position of X, rain is falling through a layer of nimbo-stratus cloud. In all probability this rain falls from a cumuliform cloud mass which cannot be seen from ground level, as it is obscured by the lower clouds. With the passage of the warm front and the arrival of the warm sector at X rain ceases, temperatures increase, and stable strato-cumulus clouds prevail. Barometric readings show a very definite fall of pressure, and the wind direction changes from easterly to westerly or south-westerly. Advection fog is common in winter as the warm front passes over the point of observation. The warm sector is characterized by stable air which is reflected in the poor visibility caused by haze from dust particles which hang in the air, below an inversion level. However, steady rainfall may also accompany the passage of the warm sector. The moist tropical air provides much orographic rainfall in the western uplands of Britain.

Weather at a Cold Front

As with the warm front, a succession of clouds and precipitation may be observed. Reference to Fig. 185 will show that at the cold front

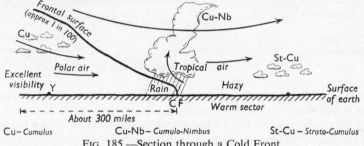

FIG. 185.—Section through a Cold Front

(CF) the polar air advances from behind the front from the direction of Y and causes a rather more abrupt uplift of the tropical air than occurs at the warm front. Even so, the gradient of the frontal surface in the troposphere is rarely much steeper than the gradient of the warm front mentioned on p. 209. The result of this relatively more steeply inclined front is to produce very high, towering, cumulo-nimbus clouds which are more common at a cold front than at a warm front. The rain belt at a cold front is quite often relatively restricted in extent, but is characterized by violent convective showers and not infrequently by thunderstorms. Wind direction changes with the passage of the front from south-west to north-west, and there are accompanying changes of temperature and pressure, the former showing a decrease and the latter an increase. Once the cold front has passed an observer at ground level it is usually followed in a relatively short time by clearance of the sky, apart from the few isolated cumulus clouds which float in the cold air. High above this cloud level tropical air flows eastwards and is largely devoid of cloud. It has been mentioned that the rain belt is restricted in area, but there are somewhat special features attached to this precipitation zone. Quite frequently the speed of advance of the polar air behind the front is sufficiently great to cause a section of the base of the front to appear as shown in Fig. 186. This may result in the overturning of air ahead of the front, and extremely violent upthrusts of warm air may occur, and these are shown by a line of cloud across the sky which rather appropriately appears to be rolling over. Sometimes this over-

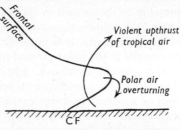

FIG. 186.—Overturning at a Cold Front

turning may be repeated several times during the passage of the cold front in a very short period. Each occurrence is associated with squally showers and gusts of wind, which are reversed in direction at intervals of a few minutes. If the slope of the cold front from the ground upwards is not very steep the rain belt may be quite extensive, since the uplift of warm air is more gradual and conditions resemble more closely those which have been described at a warm front. In passing, it may be noticed that it is rare for both fronts to show equal activity in this respect in most depressions. If the disturbance has a cold-front rain zone which is unusually extensive, then more often than not the warm front is rather less rainy.

Observation of the Passage of a Depression

In Britain the weather changes indicated at the two types of front are fairly common. It is possible to observe these changes, although the picture which has been drawn may not be seen in full until very many observations have been made. Unfortunately, it is not easy to see the whole sequence in one period, for the passage of a depression takes several hours, and in many cases some of these hours will be nocturnal. The speed of eastward movement of most depressions cannot be determined accurately, but it is unlikely to exceed 30 m.p.h. As shown in Fig. 184, the high forward cirrus clouds may be five miles above ground level at X, and at least 400–500 miles ahead of F, the warm front. Thus the gradual overclouding and the arrival of the warm front is likely to take more than twelve hours from the time of the first appearance of high cloud. The warm sector may take several more hours in which to pass, although this period will be reduced if the observer lies in the track of the junction of the warm and cold fronts or relatively near to this track. Then, after the cold front has passed the point of observation there will be several hours before the final stage of the disturbance arrives. Changes of wind direction from south-east or east in the early stages of the depression to south-west in the warm sector and to west or north-west behind the cold front, accompanied by varying temperatures, may be noted quite easily. It is possible to observe the lower clouds moving westwards in the polar air and to see higher clouds passing eastwards in the tropical air above. The passage of the warm front may be detected without reference to temperature records; the change from cool or moderately warm easterly air streams to mild or even quite warm south-westerly air streams with high humidity creates a sense of mugginess which may take place at any time in the day or night, and the physiological reactions of the body are sufficient to mark the passage of a front. Almost invariably these periods of muggy weather in the warm sector are followed by the chilliness of the polar air stream that advances behind the cold front of the passing depression. It is possible for students to practise sky reading by daytime in relation to the foregoing account.

The weather changes experienced with the passage of a depression may be different from those outlined above. If the track of the centre of the depression lies on the equatorial side of the observer, then only polar air will be encountered at ground level. The frontal surface between polar and tropical air is entirely elevated, but there may be some contrast between the forward and the rear masses of polar air, so that with the eastward progression of the depression there will be temperature changes and a decline of pressure near the centre. Precipitation is from similar clouds to those at a warm front although they are well aloft and, since polar air lies below these clouds, there is not infrequently sleet or snow.

Another possible weather sequence is that of the occluded, or partially occluded, depression, which is somewhat like the case just described. The rain belt is more extensive, and this is particularly true of a partially occluded depression where the rain belts of the warm and cold fronts have merged into one large zone of precipitation.

Depressions in each Hemisphere

The establishment of a depression, as has been shown, is dependent upon the meeting of two air masses of different quality. In the northern hemisphere, where the unequal distribution of land and sea leads to the rather special development of both continental and marine features, there is a number of recognizable fronts at which cyclogenesis (i.e. the formation of cyclones) may occur. In the southern hemisphere the distribution of land and sea is such that there is a predominance of sea areas which results in a more simple pattern of fronts, and there is only one major zone of cyclogenesis extending in a belt across the southern world in middle latitudes. Only the southern fringes of the continents of the south are affected by depressions, and the rain which is so frequently associated with the west winds in the Roaring Forties (between 40° and 50° S.) is largely cyclonic in origin. One distinction should be made between depressions in the two hemispheres. The distinction is in the movement of the winds about the centre of the depressions; in the north it is anticlockwise, whereas in the south it is clockwise (see Fig. 161, p. 185).

Families of Depressions

It is quite common for the process of cyclogenesis to be repeated on the undisturbed polar front which trails behind an occluding depression. There may be a sequence of such events giving a family of four or five depressions, each being separated by thrusts of polar air which penetrate into successively lower latitudes on each occasion. Each surge of polar air produces a tongue of high pressure, with cool, if not cold, weather and, generally, clear skies. The family ends with an eruption of polar air into very low latitudes, and before another group of depressions can form there must be a complete reformation of the polar front with an advance of tropical air into high latitudes.

Secondary Depressions

A small disturbance may arise within the original or primary depression. These are secondaries and are involved in the cyclonic swirl, moving anticlockwise around the primary in the northern hemisphere and often developing at the expense of the latter. Secondaries may arise on cold fronts of occluding depressions or over locally heated land. They give more disturbed weather than the primary; in summer they are associated with thunderstorms. On isobaric charts they may appear as a bulge, with somewhat less closely spaced isobars at one part of the primary, or they may possess their own system of closed isobars.

Troughs of Low Pressure (V-shaped Depressions)

On either side of fronts of all kinds there is a discontinuity in the trend of the isobars as shown on weather maps. In the past, when this feature was very pronounced it was known as a V-shaped depression from the pattern of the isobars, but today it is called a trough of low pressure (Fig. 187). The discontinuity is most noticeable at cold fronts at whose passage there is an almost complete reversal of the winds and a very considerable fall of temperature. The trough is accompanied by very heavy rain and violent squally showers with thunder and lightning. There is usually a very rapid clearance in the weather after the trough has passed.

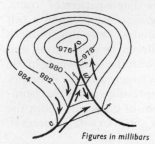

Figures in millibars

FIG. 187.—A Trough of Low Pressure or a V-shaped Depression (Northern Hemisphere)

$w f$ = warm front.
$c f$ = cold front.
$o f$ = occluded front.

Anticyclones

Anticyclones are systems of high pressure which may be permanent, as in the case of the highs over the subtropical oceans or the semi permanent continental highs which cover Eurasia, and to a less extent North America, in winter. On the other hand many anticyclones in temperate latitudes are relatively of short duration, and represent either extensions of cold polar air behind depressions or poleward extensions of subtropical highs over the oceans and sometimes over neighbouring land areas as well. In all cases they are defined on weather maps by enclosed isobars, but the spacing of the isobars is usually much greater than is typical of the depression pattern. Winds are therefore gentle compared with those in a depression. The central area of an anticyclone is one where air subsides and diverges as an outward flow at the surface of the earth from the focus of high pressure (Fig. 188). The circulation in an anticyclone is the reverse of that of a depression, in other words, in the northern hemisphere the winds move in a clockwise swirl, whilst in the southern world the swirl is anticlockwise.

Anticyclones may be divided into two classes, warm and cold. The former type has warm air in its lower layers and this is relatively light,

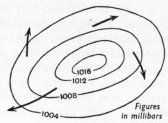

Figures in millibars

FIG. 188.—Anticyclone (Northern Hemisphere)

so that there is an excess of air at higher levels in the troposphere which produces the high pressure that is so characteristic of the anticyclone. Cold anticyclones have dense air in the lower layers. The permanent large-scale oceanic highs in subtropical latitudes are warm anticyclones, whereas the polar highs and the continental winter anticyclones of the northern land masses are of the cold class. Less permanent warm and cold anticyclones are produced by temporary extensions of tropical and polar air, respectively, into the middle latitudes. The former may be fairly persistent, but the latter are often mere interludes of fair weather between the passage of individual depressions, although they may be fairly long-lived in winter.

In most anticyclones there is a marked inversion of temperature not far above the ground, and it is at this level that some cloud may form, producing anticyclonic gloom, with drizzle. Above this inversion the troposphere is very clear. In winter, anticyclones lying over land areas give very cold weather with clear skies which induce radiation fog. In addition, advection fog is formed when the warmer air on the equatorward margins of an anticyclone passes from the sea to the relatively cold land. Frost is also very common owing to the great loss of heat from the surface of the earth by radiation. In summer, warm anticyclones give fine, warm, and dry weather, although, over the sea, fog may be common.

Cols

Two tongues of high pressure, one usually of polar origin and the other of subtropical origin, may separate two zones of low pressure, or in some instances two adjacent depressions of a family. The isobaric pattern (Fig. 189) resembles that of the topographic col on contour maps, hence the name. Rather varied weather is characteristic of a col, and winds may blow from various directions. With the meeting of air from different sources, fronts may form in the col, and often there is a succession of depressions passing between the tongues of high pressure.

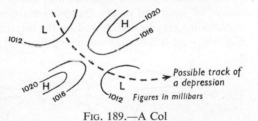

FIG. 189.—A Col

Tropical Cyclones

Up to the present the cyclones which have been described have occurred in middle latitudes. Cyclones do arise in the tropics and have similar clockwise and anticlockwise circulations according to the hemisphere, but, in general, their progression is westwards and not eastwards. The deflecting force, which is set up by the rotation of the earth, is absent at the Equator, and cyclogenesis is well-nigh impossible there. Tropical cyclones develop on either side of the Equator, usually between 10° and 20° of latitude, north and south, although they may be experienced nearer the Equator.

Tropical cyclones form over oceanic areas with strong instability where the warm humid air rises and produces heavy rainfall. The

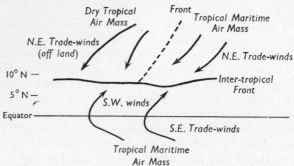

FIG. 190.—A Triple Point on the Intertropical Front
(*After Pack*)

condensation which precedes this rainfall is instrumental in creating heat and, therefore, energy which maintains such a disturbance. It is possible that a critical contrast may occur when a tropical air mass moves off land and meets a tropical maritime air mass. Sometimes a triple point may be recognized on the intertropical front as shown by Fig. 190, and it is at such a convergence that a tropical cyclone is likely

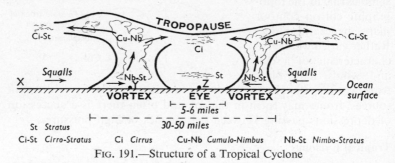

FIG. 191.—Structure of a Tropical Cyclone

to develop. It is rare to find as in temperate depressions a system of fronts within these cyclones. Other opinions suggest that an influx of cold air at high levels in the Doldrums belt is responsible for the formation of a tropical cyclone.

The general structure of such a cyclone is shown in Fig. 191. It appears on weather maps with isobars which are very nearly circular.

Ahead at X, outside the cyclone, there is almost dead calm, with oppressive heat and great humidity; there is a low ground swell at sea which may be detected, with unusually high tides on coasts, as far as 800 miles ahead of the storm. As Y the vortex of the cyclone approaches X, there is a distinct lowering of the air pressure and an onset of squally weather, with a thickening of the cloud cover which may gradually extend upwards from the sea surface to the tropopause. The vortex, at Y, may be fifteen to twenty miles across and may take as long as four hours to pass an observer. During this period winds are violent, commonly reaching speeds of 150 m.p.h. The vortex is accompanied by thunderstorms with incessant lightning. The thunder is rarely heard above the noise of the wind and heavy rain, which comes from a dense mass of cumulo-nimbus cloud. Pressure of the air falls by as much as 50, and even 100, millibars near the inner edge of the vortex.

After the vortex comes a sudden change to the 'eye' or inner calm area at Z, which may take about two hours to pass, but usually much less time, often a matter of ten to fifteen minutes. In the 'eye', pressure remains very low but fluctuates with a range of some 10–15 millibars, which is probably a reflection of the variation that takes place in the area of the 'eye', which seems to contract and dilate. The period of calm is followed by the violence of the other side of the vortex, but here the wind direction is the reverse of that prevailing in the earlier vortex period. Gradually air pressure rises as the storm recedes.

The approach of such a storm is marked by the massive pile of convective cloud which is seen as far as 150 miles away by day, but at night it may be detected from a distance of 250 miles by the vivid lightning display. These cyclones are very destructive, and if they approach closely to a shoreline they tend to heap up the sea and inundate coastal towns.

Their progression is initially westwards over oceanic surfaces, but they show a remarkable recurvature by swinging round the western fringes of one of the subtropical high-pressure cells and away from the Equator, so that very often they finish their existence by entering the westerly wind belt on the poleward side of the high-pressure area. The curve followed by the storm is approximately a hyperbola. It is on recurvature that the cyclone may come into close proximity with land areas. If cyclones of this type pass onshore, thus leaving the ocean source of humid air, their intensity is rapidly reduced and, assisted by frictional effects of the irregular land surface, they are soon disintegrated. After recurvature and further progression over an ocean, the violence of the storm is much less, it becomes elongated, probably because of the development of a front, and it enters the westerly wind belt to give heavy frontal rainfall when it meets colder, polar air in temperate latitudes.

Tropical cyclones occur usually in August and early September in the northern hemisphere and about February and March in the south. These periods of the year coincide with the most northerly and southerly positions, respectively, of the intertropical front which may be relevant. Areas which are most regularly visited by these storms are the Caribbean zone (hurricanes), South-east Asia and in particular the south coast of China (typhoons), and the Guinea coastlands of Africa. In the southern hemisphere they are relatively infrequent, but the east coasts of most continents may be visited by tropical cyclones, as is shown by the map of their distribution (Fig. 229). The freedom from these cyclones which is enjoyed by Brazil seems to be related to the complete absence of the intertropical front over the Atlantic Ocean south of the Equator.

Special Winds Associated with Depressions and Anticyclones

These winds fall into three groups:
1. Winds associated with the passage of the fronts of depressions.
2. Cold winds, such as the bora and the mistral.
3. Warm winds of the föhn type.

Of the first group it may be said that the winds are typical of all depressions where tropical air streams of the warm sector alternate with polar air streams. In some areas, particularly in the warm temperate latitudes, they have local names. In the Mediterranean Sea basin warm southerly winds blow from the Sahara and are known as the leveche[1] in Spain, sirocco[2] in Italy, and khamsin[3] in Egypt. The cold bora and mistral from the north are discussed below. In California a warm desiccating wind called the Santa Ana[4] is drawn north by depressions in somewhat higher latitudes, whilst cold northers flow southwards over Texas and the Gulf States in the rear of depressions. The cold blizzard of Canada and the buran of Siberia are of similar origin. In Argentina, eastward-moving depressions may bring the warm zonda in front and the cold pampero in the rear, whilst in South-east Australia there are warm brickfielders and cold southerly bursters associated with troughs of low pressure (see Fig. 232).

Mistral and Bora

These are winds of restricted extent, the mistral blowing in Provence and Languedoc, the bora on the north Adriatic shores in

[1] Leveche—wind from the south.
[2] Sirocco—from the Arabic 'Sharq', meaning the east.
[3] Khamsin—meaning 'fifty', since the wind was said to blow for fifty days.
[4] Santa Ana—the name of the valley down which the wind blows.

Yugoslavia.[1] Although related to the rear of winter depressions moving eastwards through the Mediterranean Sea basin, they are also dependent on cold areas with high pressure lying over the Central Plateau or the Maritime Alps of France and over the Dinaric Alps respectively. Cold air descending from the mountains is warmed adiabatically, and yet arrives on the Mediterranean coast of each area as a very cold local wind.

It is possible that these are katabatic winds. Such winds are caused by nocturnal radiation which chills the mountain air and thus increases its density so that it flows downhill. Katabatic winds are responsible on winter nights for pools of cold frosty air which collect over valley floors with a temperature inversion above. In daytime anabatic winds are induced by the heating of the lower slopes in mountain valleys, when the air in contact with these surfaces is warmed sufficiently for it to flow uphill, being displaced by cold air from above.

Föhn Winds

The föhn wind belongs properly to the northern slopes of the Swiss and Austrian Alps. Similar winds are found elsewhere, such as the chinook on the eastern slopes of the Rocky Mountains between Alberta and Colorado, the nor'wester from the Southern Alps of New Zealand, and the berg wind in the Cape region of South Africa.

Föhn winds blow down mountain slopes into adjoining lowlands and valleys and bring very warm and dry air. Local increases of temperature are remarkable (particularly in winter) and of the order

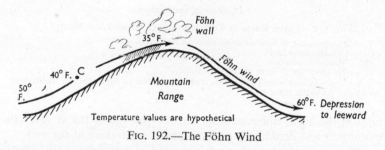

FIG. 192.—The Föhn Wind

of 40° F. in twenty-four hours, although an increase of 30° F. in an hour has been recorded. The onset of the wind is shown on thermograms by fluctuating temperature, but after three or four hours a steady high temperature is recorded which persists for several hours

[1] Mistral = masterly.   Bora = north.

at least. Fig. 192 shows the difference in temperature values at corresponding levels on either side of the mountain area.

The wind has been explained in this way. A depression to leeward of the mountains draws air up the windward slopes and at C condensation occurs with the release of heat. Further ascent is accompanied by adiabatic cooling at the more gentle lapse rate for saturated air. The mass of cloud which accumulates at the crest of the range is known as the föhn wall (or chinook arch). The air descends on the leeward slopes as dry air, and throughout most of the descent it is warmed adiabatically at the dry lapse rate which is relatively rapid. The process is illustrated graphically in Fig. 193, which shows the lapse

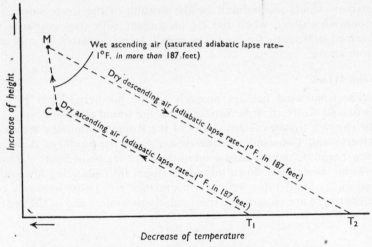

FIG. 193.—Lapse Rates at Different Stages of the Föhn Wind

T_1 = Temperature at base of mountain on windward side.
C = Condensation level.
M = Mountain summit.
T_2 = Temperature at base of leeward slope.
    Note.—No cloud is brought down from M, so that air descends the whole way with dry adiabatic lapse rate.

rates in the air at different stages in the passage of the air over the mountains and demonstrates the higher temperature at the end.

Today this explanation is not regarded as completely satisfactory, because the temperature contrasts are greater than can be expected if the process described above operates alone. Furthermore, condensation, with the formation of clouds, does not always take place, so that the air remains 'dry' throughout the process. Recent studies of the chinook suggest that the föhn effect is primarily caused by turbulence in the air over the leeward slopes. Fig. 194 indicates that an air stream on crossing mountains causes great eddies on the lower leeward slopes, which in turn create a forced descent of air. Some of the air which is thus forced to descend is of tropical origin, and will be relatively warm and becomes warmer by adiabatic compression. Thus

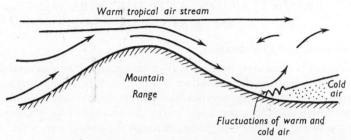

Warm tropical air stream

Mountain Range

Cold air

Fluctuations of warm and cold air

FIG. 194.—Modern Explanation of the Chinook Wind
(*After Willett*)

its temperature is raised to the high values which are in accord with those registered at the mountain foot. As the warm air replaces cold air there is a series of surges of cold and warm air which is reflected in the alternation of temperature that has been mentioned as typical of the onset of the föhn wind. At the same time the lower layers of the tropical air suffer the transformations shown in Figs. 192 and 193, so that the earlier explanation is not completely invalidated.

CHAPTER XIV

PRECIPITATION AND DISTRIBUTION
OF RAINFALL

Measurement of Precipitation

Rainfall is measured by means of a rain gauge. The essential features of this instrument are shown in Fig. 195. It is a copper cylinder in two parts; the upper section has a deep rim around a funnel of metal which leads into the lower part. The rain falls over the area of the top of the cylinder and passes through the funnel into a jar. The

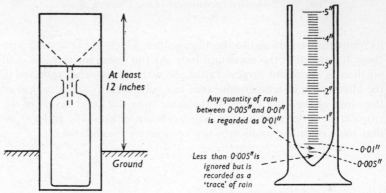

FIG. 195.—The Rain Gauge

At least 12 inches

Ground

Any quantity of rain between 0·005" and 0·01" is regarded as 0·01"

Less than 0·005" is ignored but is recorded as a 'trace' of rain

0·01"
0·005"

FIG. 196.—The Measuring Cylinder for the Rain Gauge

area of the cross-section of this jar is less than the collecting area at the top of the cylinder. The rain-water which is thus collected has to be poured from the jar into a measuring glass, which is graduated so as to read directly the depth of rain that has fallen over an area equivalent to that of the top of the cylinder. Rain is recorded in inches or fractions of an inch, since on some days the amount is very slight. The measuring glass has a cross-section which tapers very much at the base (Fig. 196), and thus very small amounts of rain may be recorded. The smallest amount that can be registered is $\frac{1}{100}$ inch, and any quantity which is less than this is ignored. The gauge is placed with its upper rim a foot above ground level in order to avoid increasing the amount collected through splashing into the cylinder.

When snow is precipitated, the measurement is difficult, since it must be melted before the equivalent depth of water can be measured. When the snow is deep or when it has drifted an accurate estimate is very difficult to make.

As with temperature and pressure distributions on world maps, rainfall may be shown by lines similar to isotherms and isobars. These lines are known as isohyets, and are drawn from mean totals of the recorded rainfall over a given period. No reduction for height above sea-level is necessary. Since rainfall may fluctuate from year to year, the mean values must be regarded with care. This is particularly true in areas with scanty rainfall in warm or hot regions. It will be seen from Fig. 197 that the mean annual rainfall of a station with a very dry climate is 1·1 inches. This value is given by the average of 35 totals, most of which are less than 1·1 inches. It is a few 'wet' years that produce a mean value which is not typical as far as most years are concerned. This example is applicable to monthly means as well as annual values, and it also illustrates what is true of many stations in arid or semi-arid areas and of some areas which have quite abundant annual rainfall, where the rainfall may be twice or thrice the mean in certain years and very much less than the mean in others.

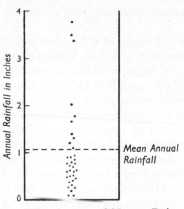

FIG. 197.—Average of 35 years Rainfall in Dry Regions

Each dot represents total for one year.

Types of Rainfall

It has been shown that rainfall is dependent on a series of events which are usually initiated by air rising from ground level into the troposphere (see p. 189). There are three types of rainfall, each of which is distinguished by the chief cause of such uplift. They are:

1. Convectional rainfall.
2. Frontal or cyclonic rainfall.
3. Relief or orographic rainfall.

Convectional currents have been illustrated in Fig. 165, and all that need be added here is that either convection is great enough to overcome the stability of the troposphere and air is forced to ascend to great heights, or as is more usual, local heating produces the initial impetus which sets air in motion, when its further uplift is determined by instability. Convectional rainfall is thus typical of hot seasons or of hot periods in a day, for example, a summer afternoon. The rain is usually very heavy, in the form of short showers, not infrequently accompanied by thunderstorms. It is typical of equatorial and tropical areas at all seasons, and of continental interiors in middle and high latitudes in the summer season.

The second type of rainfall has been discussed fully in the previous chapter, where the characteristics of the various types of fronts in a . depression were described. In this case there is uplift of warm and light air by a wedge of colder and heavier air which, if carried far enough upwards, will give the condensation features that are preliminary to rainfall.

Relief rainfall is caused by the forced ascent of air by features of high relief of the land which lie in the path of winds. Wherever onshore winds from oceans are humid and warm in relation to the land barrier on to which they are blowing, the rainfall of this type will be particularly heavy. The level of maximum precipitation of this kind is usually not far above the level at which precipitation begins on the mountain-side. Farther up the totals of rainfall are somewhat less. Whilst relief may produce heavy rainfall, its effect is more often than not an added factor to some primary cause of rainfall, such as frontal uplift or the heating of the ground surface, with the development of convection within the upland area. Much orographic rainfall is associated with the air of tropical origin in the warm sectors of depressions.

Thunderstorms

The cumulo-nimbus clouds (Fig. 198), which usually develop before a thunderstorm, are produced, largely, by very strong convection currents, due more often than not to excessive ground heating. The atmosphere has at all times particles which carry charges of electricity, both positive and negative. Between these oppositely charged particles there exists an electric potential gradient. If the normal gradient is increased to very great proportions, then very large sparks may leap across spaces in the atmosphere and give lightning. The accumulation of the great positive and negative charges of electricity which are necessary to produce a flash of lightning has been explained in various ways, but there is no completely satisfactory account of the process. Simpson's theory is both simple and reasonably acceptable. It turns largely on the fact that a very large drop of water will disintegrate into relatively small droplets as it falls through an updraught of air; some of these will be negatively charged whilst others are positively

Clark Collection, Royal Meteorological Society

FIG. 198.—Cumulo-nimbus Cloud

The cloud shows rain falling from its base.

charged. The constant repetition of this disintegration within a cumulo-nimbus cloud produces an upper layer of negatively charged drop-lets and a lower layer of positively charged drops. Lightning may flash either between the upper and lower layers of the cloud or between the positively charged cloud base and the surface of the earth below, which is always negatively charged (Fig. 199).

Some of the rain-drops which fall from a cloud must evaporate and cause some cooling, since heat is extracted from the surrounding air when drops of water evaporate. The farther the rain-drops fall the more extensive will be the cooling by this process, and at ground levels the spread of cooled air will interfere with convection and this will eventually weaken and finally destroy the thunderstorm. Hail, formed by large rain-drops being carried up to very high levels in the cloud and there freezing through direct condensation of ice from supercooled water drops, is a common feature of the precipitation from a thundercloud. The size of hailstones is often so great that it is estimated that, within these clouds, there are ascending currents whose speeds are of the order of 100 m.p.h. In the equatorial latitudes thunderstorms are very common, chiefly because of the large mois-

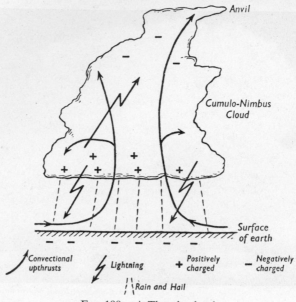

FIG. 199.—A Thundercloud

ture content of the air which produces considerable instability that reaches to very high levels in the zones of air-mass convergence.

Distribution and Seasonal Incidence of Rainfall in the World

Having discussed the distribution of temperature, pressure and winds, and air masses in the world in previous chapters, it is possible to show their interrelations which result in the distribution of rainfall over the earth's surface. High temperatures prevailing during most of the year or in a hot season will produce convectional rainfall, whilst pressure distributions are responsible for rain when winds move on-shore or meet high relief. Furthermore, pressure distributions control the convergence of air masses which may produce frontal rainfall.

It is necessary to know where there is abundant or deficient rainfall, but it is perhaps more important to know the seasonal incidence of the rain. Rainfall may occur in all months of the year, in certain months only, or no rain may fall in any month of the year. This seasonal incidence of rainfall is known as the rainfall régime.

In discussing the world distribution of rainfall, including its seasonal incidence, consideration must be paid to two factors other than those mentioned above. Firstly, the migration of the overhead sun between the tropics during the year is reflected in a similar but less extensive swing of the pressure and wind belts (as shown in Fig. 200) both northwards and southwards. Secondly, there is the effect of large land masses creating special controls which are known as continental. Whilst such an effect is found mainly in the northern continents, it is of some significance in tropical areas generally, but also, more particularly, where monsoonal controls are operative in low latitudes.

The apparent movement of the overhead sun between the tropics has an amplitude of 47° of latitude, whereas the main pressure and wind belts move north and south of their mean positions with a range of not more than fifteen degrees of latitude over the oceans, but over land areas there is a somewhat greater shift of these belts. Referring to Fig. 158 it will be seen that there must arise from these migrations three zones in each hemisphere where, during the year, there is a change in the predominant influence. Thus, immediately north and south of the equatorial zone there are regions which are alternately under trade-wind influences and the Doldrums belt. About latitudes 30° and 40° there are zones which have trade-wind influences at one time, and westerly winds with their associated depressions at others, as a result of the movement of the subtropical high-pressure belts. In high latitudes are areas which tend to have polar influences in winter and westerly winds in summer, although the interaction of polar and tropical air masses produces depressions which upset and obscure the simple pattern.

Fig. 200 ohows the essential effect of such movements by giving the relative positions of the two subtropical liigh-preccure cells over the oceans, and the Doldrums low-pressure belt, together with their associated winds at the equinoxes and at the two solstices (i.e. when the sun is overhead at the Equator and at either tropic, respectively). Two 'ideal' continents are shown between 60° N. and 60° S., and are separated by an ocean, so that amongst other features shown by the diagrams there is the contrast between the eastern and western margins of the land masses, with onshore winds on one margin matched by offshore winds on the other. Unfortunately, the diagrams present a generalized pattern which must be modified for a complete account of world rainfall. Nevertheless, they serve as a basis for some of the climatological studies which follow in other chapters. Several areas are indicated on each continent by letters which appear on all three parts of Fig. 200, and reference will be made to these in turn. The shaded areas mark regions of adequate, if not abundant, rainfall.

Let us consider first the areas A and B. These are the equatorial zones with permanent low pressure in the Doldrums belt, and are

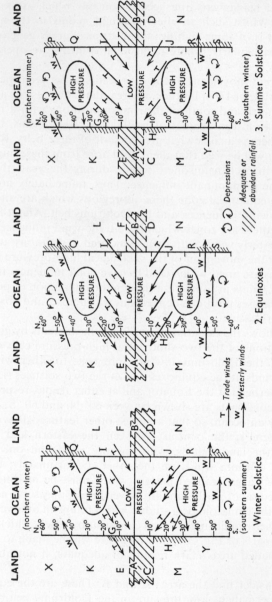

Fig. 200.—Seasonal Movements of Pressure, Wind, and Rainfall Belts

characterized by fairly constant convectional rainfall. At the equinoxes these zones should have an equal extension north and south of the Equator, but at the winter solstice both A and B have shifted to the south, so that areas C and D are now covered by the convection rain-belt whereas at the equinoxes they were relatively dry. At the summer solstice the reverse has happened and now regions E and F have rain, although previously at the equinoxes and the winter solstice they were relatively dry (during these relatively dry periods some rain will be received from trade-wind sources). Thus, as a result of this seasonal migration of the low-pressure belt, there is a zone of well-distributed rainfall near the Equator which merges northwards and southwards into zones that receive convectional rains in the period of the year when the sun is overhead in the particular hemisphere, or in the summer season. There is a gradual decrease in the annual total of rainfall as one moves polewards through these zones.

It will be seen that the trade winds migrate and cover areas of varying size in each hemisphere with the shift of the oceanic cells of high pressure, but the most significant fact is the difference between the eastern and western margins of the two continents in low latitudes, the former margins receiving rain, whilst the latter are dry. At G and H onshore trade winds will give rain, particularly where the relief is high. It should be noted by comparison of all three parts of Fig. 200 that these rain-belts on the east coasts have poleward limits which vary with the season. At I and J offshore trade winds prevail at all seasons and these are drought zones, and they are linked respectively to L and N which are also dry at all times of the year. L and N correspond to K and M, since all four regions are poleward of the areas affected by the migration of the tropical convectional rain-belt. Also these regions are inland and therefore remote from any rain brought by the trade winds, and are largely desert in character.

Farther north and south the major influence is that of the westerly wind circulation, with its associated depressions. The contrast between the two continental margins is again very obvious, but in this case the western margins receive rain. With the migration of the pressure and wind belts the areas P and S are sufficiently high in latitude to receive rain at all seasons, but Q and R are less fortunate in that they lose the rain-bearing winds and depressions from the west at the summer and winter solstices respectively, or, in other words, they are dry in summer but have rain in the winter season.

Regions X and Y appear to be like areas I and J, completely dry at all seasons, but this is not true, as will be seen below. We have been dealing with ideal conditions, and no account has been taken of the influence of continentality which is so important in each of the great northern land masses. It was noted (p. 168) that the mean isotherms in January and July over North America and Eurasia were very irregular as compared with those in other parts of the world maps. The chief features to be recalled at this point are:

1. The southward bend of the isotherms over these continents in winter which indicates that the east and west margins are much warmer than the interiors in corresponding latitudes. This is particularly true of the west margins, where the isotherms have very nearly a north–south trend.

2. In summer the reverse is largely true, and the interiors are very much warmer than the margins in similar latitudes.

This contrast is reflected in the world pressure maps, where it is seen that over the northern continental interiors high-pressure air which is dominant in winter is replaced by low pressure in summer.

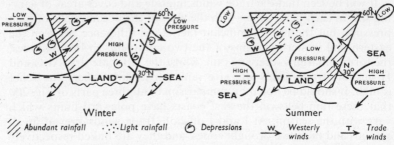

Winter   Summer

/// Abundant rainfall   ·∵· Light rainfall   ⑥ Depressions   W→ Westerly winds   T→ Trade winds

FIG. 201.—The Effect of Continentality in Northern Land Masses

The result of these contrasts may be shown diagrammatically as in Fig. 201 which is derived from Fig. 160 (see p. 184).

From the above diagram it will be seen that the northern continents in middle and high latitudes may be divided into three main areas:

1. The western margins which receive rain from the westerly winds and depressions at all seasons, but with greater certainty in winter towards the north, when the buffer of the continental high-pressure region causes most depressions to be deflected polewards.

2. Eastwards of these margins there is an interior zone which is effectively isolated from rain-bearing winds in winter by high pressure, but is open to westerly influences in summer. With the additional convectional rainfall produced by great heat, this is unquestionably the rainy season of the year.

3. The eastern margins are often quite exceptional in that they are under monsoonal controls. Here, as shown by the two diagrams, there is an almost complete reversal of the wind directions between

the seasons. Clearly, summer is the period of onshore winds and abundant rainfall. It will be noticed that the summer winds are completely controlled by the low-pressure area of the interior, so that they are deflected very much from the normal directions in these latitudes. In winter, outblowing winds from the continental high-pressure area replace the inblowing winds of summer and, in general, there is little rain at this season, but depressions bring some winter rainfall. Thus it will be seen that region X is not dry as indicated in Fig. 200. In corresponding latitudes in the southern hemisphere the lands are too narrow to develop this continentality, and region S in Fig. 200 represents Southern Chile, New Zealand, or Tasmania, whereas region Y resembles very closely the Patagonian part of Argentina with low rainfall. The monsoonal régime is probably typical of but one region of the world, namely, the south and east margins of Asia, with the East Indies and Northern Australia, which are represented by G and H in Fig. 200, but from what has just been said these tropical areas with their predominant summer rainfall merge northwards with the eastern margins in higher northern latitudes where in Asia, at least, there are monsoonal controls.

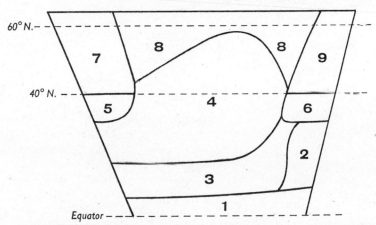

Fig. 202.—Distribution of Rainfall and its Seasonal Incidence on an Idealized Northern Continent.

1. Rain at all seasons (convectional).
2. Rain at all seasons (convectional and from the trade winds, also monsoonal).
3. Rain in summer season (convectional—one or two periods).
4. Scanty rainfall or none.
5. Rain in winter only (westerly winds and depressions).
6. Rain mainly in summer (monsoonal winds and depressions).
7. Rain at all seasons (westerly winds and depressions).
8. Rain mainly in summer, decreasing in amount to the east (mainly convectional with some from depressions).
9. Rain mainly in summer, in places monsoonal (some in winter from depressions).

The great width of each northern continent makes it very unlikely that rain-bearing winds and depressions can bring much rain to the remote parts of the interiors in summer. Thus there are practically rainless areas within the interiors which may be regarded as poleward extensions of the desert regions of low latitudes (areas K and L in Fig. 200). This is particularly the case in Eurasia, but less true of North America. The resultant distribution of rainfall, with its seasonal incidence, on an idealized northern continent is given in Fig. 202, and within certain limits this may be reversed for a southern continent.

Part IV

PLANTS AND SOILS

CHAPTER XV

VEGETATION

Controls of Vegetation

The vegetation of a particular region may be regarded as the community of plants, which includes trees and grasses, that covers the area and gives it a distinct character. From the geographer's standpoint the types of vegetation depend on the appearance of the plant cover, e.g. forest and woodland, grassland and desert. There are no hard-and-fast limits to each of the types, rather do they merge imperceptibly with one another through a series of intermediate types of vegetation. The gradual change from one distinct type to another is largely the result of the primary control of rainfall, and more particularly its distribution throughout the year. However, some differentiation of vegetation is dependent upon the temperature, such as the distinction between tropical rain forests and northern coniferous forests, and locally the soil may be a factor which must be considered.

Before examining in detail the various types of vegetation, it is necessary to consider the requirements of most plants during their growth. When it is growing, a plant must have a supply of water which carries dissolved substances from the soil through the roots and stem to its leaves. In each leaf these substances, brought from the ground by this transpiration stream, combine with other substances made by the action of light in photosynthesis and form complex plant foods which are carried back to the cells within the root and stem or trunk, where they are either stored or used to promote further growth. This complex process depends very much upon the constant movement of solutions within the plant. Through the stomata, which are minute pores in the leaves, water vapour is transpired into the air. Broad-leaved trees lose much water in this way during the day and, since transpiration depends upon the same factors as evaporation, this loss is increased by winds and by a dry atmosphere.

Obviously, a growing plant must have an adequate supply of moisture to maintain this transpiration stream from roots to the leaves. However, the transpiration stream is probably arrested, or at least slowed down, by cold, and it is accepted that, in general, plant growth is inhibited when mean monthly temperatures remain below 43° F. (the zero of plant growth is often taken as 42° F.). Nevertheless, plants and their seeds may exist, without growth, at much lower temperatures, and some are capable of living with temperatures well below freezing-point. It is reasonable to assume that, in the majority of cases, temperatures must be some 10° F. above freezing-point for anything like active growth to occur.

With the onset of cold it is likely that the transpiration stream slackens despite the fact that there may be available water in the soil. However, low temperatures also slow down photosynthesis and other vital processes in the plant. This means that the plant ceases to produce food which leads to a fall of the leaves and a rest in growth. Not until fresh leaves have been developed can the transpiration stream come into operation again and growth start afresh.

When plants remain evergreen during such cold seasons, the leaves have small surface areas (as in the case of the needle leaves of the coniferous trees) or they have thickened cuticles (as in the case of the holly) to reduce the rate of transpiration. It is sometimes necessary for trees to shed their leaves because of drought instead of cold, but again the probable purpose of the leaf fall is to prevent damage to the plant through transpiration at such critical periods when the soil moisture is not replenished by rainfall. Few areas of the world enjoy optimum conditions for plant growth, with high temperatures and abundant rainfall throughout the year, and many plants are therefore adapted to the climate of the region in which they grow.

The relation between climate and vegetation is of the utmost significance. Given suitable growing temperatures, the most important factor in plant growth is the incidence of rainfall. In areas other than those with abundant and well-distributed rainfall, the incidence and the length of a dry season is of great importance in determining the vegetation. Normally, plant growth is most vigorous at the time of the greatest heat. Amongst tropical and warm temperate areas, where mean monthly temperatures are always above 43° F., those with dry seasons (with the exception of deserts and, also, the Mediterranean areas) have rain at the time of the greatest heat. In cool temperate and cold regions the primary factor in plant growth is the length of the season with mean monthly temperatures above 43° F., but even so, the incidence of rainfall plays a very important rôle as in warmer regions.

The trees of the tropical rain forest are hygrophilous in that they require abundant rainfall at all seasons. In deserts the plants must be adapted in various ways to resist drought and are termed xerophilous. Some plants are tropophilous, being adapted to seasonal changes of rainfall and temperature, such as the deciduous trees of our latitudes which are in leaf in the moist summer and rest in their growth during the cold winter, after the leaf fall. A season of drought will also give rise to tropophilous plants. With the intervention of a dry season, forest is replaced by woodland and scrub which are tropophilous in character. Grassland flourishes where the rainfall is concentrated in the warm season of the year and is insufficient to support forest or woodland (although there is by no means a complete absence of trees). Grassland may be regarded as tropophilous vegetation since in the rest period there is usually a drought, and then it may be said to exhibit xerophilous characteristics. Deserts have rainfall at such rare intervals that only specially adapted plants may grow and the vegetation is essentially xerophilous. In other words, with increasing length of the dry season, forest passes through woodland and scrub to grassland and desert.

Xerophilous plants are often regarded as suited to dry areas because they transpire very little. It is now known that with an adequate water supply these plants may transpire as freely as many others. Whilst the stomata are reduced in size these are, in fact, relatively more numerous than in many other plants. In some cases the stomata are sunken to reduce transpiration. These plants conserve water in their cells during drought when transpiration is checked by the closing of the stomata. Various other devices, to be referred to later, help these plants to resist drought. Perhaps, above all, whilst these plants remain in a wilted state in the dry seasons they can recover without permanent damage being done to their protoplasm with the onset of another wet season.

On the other hand, the flora (or types of plants) within a forest or grassland area depends, not so much on the rainfall, as on the prevailing temperatures. Thus, the particular trees within a tropical rain forest are different from those in a coniferous forest and the grasses within the tropical grasslands are unlike those in the grasslands of high latitudes.

It will be seen that the vegetation of a region is primarily dependent upon its climatic régime, and it may be said that in the cooler regions temperature is more critical in its control than rainfall, whereas in low latitudes rainfall produces the major differentiation of vegetation. However, other factors such as soil may produce local variations, as, for instance, where forest appears on water-retentive clays in grasslands or where grasses flourish on very pervious soils in forest areas. In the remainder of the chapter the close relationship between climate and vegetation will be established.

Forest and Woodland

In forests and woodlands the predominant vegetation consists of woody plants, and of these, trees are the most highly developed.

Forest may be defined as an assemblage of trees in such close growth that their crowns touch one another. Woodland is characterized by the considerable development of shrubs which tend to isolate the taller trees. Scrub is the term applied to vegetation where trees are few in number and occur in scattered clumps, accompanied by many of the lower woody plants as well as by grasses. Only within the tropical rain forests and in the forests on the eastern margins of warm temperate regions do trees enjoy the optimum conditions for growth at all times of the year. In all other forests and woodlands there is some restricting factor which imposes a seasonal rest in the growth of the trees.

Trees are distinguished in various ways, and some confusion may arise in the application of certain terms. Probably the most important contrast lies between deciduous trees, with a seasonal leaf fall, and evergreen trees which lose their leaves continuously but never all at one time. Another distinction depends upon the shape of the leaves of the tree, some having broad leaves and others needle leaves. It is incorrect to assume that broad leaves are solely characteristic of deciduous trees, as many broad-leaved trees are evergreen. On the other hand, most coniferous trees bear needle leaves. In general, broad-leaved trees, whether deciduous or evergreen, yield hardwoods, whilst softwoods are mainly derived from coniferous trees which are commonly evergreen. There are some exceptions, however, in respect to the type of wood obtained from these various trees.

Forests may be grouped geographically as follows:

1. Tropical Forests

(a) *Tropical Rain Forests.*—These are found in regions with abundant rain at all seasons, in equatorial lowlands, and on tropical east coasts, in particular. In the Amazon basin they are known as the selvas. They are broad-leaved evergreen forests, since temperatures permit of growth at all seasons. With no rest period leaves are shed at the same time as fresh ones are developing. The forests have relatively little undergrowth, except near the margins, since the canopy of the tree crowns restricts sunlight on the floor of the forests. The trees are, however, encumbered with climbing plants such as lianas. The forests are extraordinarily rich in species of tree, and include hardwood trees, such as mahogany, rosewood, and greenheart. Mangrove forest, which flourishes in the salt waters immediately offshore from coastal rain forests and in the waters of tidal rivers, may be regarded as a subtype. Tropical rain forest is shown in Fig. 215.

(b) *Dry Forests.*—These forests belong to the tropical continental interiors where a dry season is characteristic. The forest is semi-deciduous or, more correctly, facultatively deciduous, in that some but not all of the leaves are shed, and the degree of the leaf fall is dependent upon the severity of the dry season. Only in very dry years do these forests appear leafless. This less dense forest has some undergrowth, since sunlight can penetrate fairly readily, but usually this remains evergreen. Common trees are acacias. In South-east Asia and in India comparable forests are known as monsoon forests, with

Pictorial Press

FIG. 203.—Dry Forest in Rhodesia
This is an example of dry forest with acacias.

teak and sal trees which are similarly adapted by a leaf fall to existence through a dry season. Dry forest is shown in Fig. 203.

(*c*) *Scrub and Thorn Forests.*—These appear in tropical regions with longer dry seasons than obtain in dry forest zones. The trees are almost completely deciduous, and include acacias, euphorbias, and baobabs; the latter develop a water store in their excessively swollen trunks to endure the drought. In North-east Brazil thorn forest is known as the caatinga. Scrub is shown in Fig. 223.

Dry forest and scrub are accompanied by grasses, and it is usual to find them occupying much of the areas that are often referred to as savana. Savana, strictly speaking, means a treeless plain.

2. Warm Temperate Forests

(a) *Mediterranean Broad-leaved Evergreen Forests and Scrub.*— These are of considerable interest, since they are adapted to a summer drought. The mean monthly temperatures permit active growth during the rainy winter, but the trees remain evergreen in summer, instead of becoming xerophilous in character during the drought by a leaf fall. The trees have various devices to resist the drought and these include spiny leaves, with waxy and hairy surfaces, thickened cuticles, incurved leaves with stomata turned away from the direct sunshine and, in many cases, very long tap roots to draw upon deep supplies of water, as well as very thick and deeply fissured barks. In general, the trees are widely spaced, and show but moderate vertical development, and they are usually accompanied by woody shrubs. This scrub is known by various names, such as maquis in Europe and chaparral in California. Amongst the trees are cork oaks, sweet chestnuts, and olives, together with conifers such as cedars and cypresses. Few areas have true forest although in Swanland, in Western Australia, the eucalypt[1] forest, with jarrah and karri trees, is remarkable. Maquis is shown in Fig. 230.

(b) *Broad-leaved Evergreen Forests.*—These are found in the eastern marginal areas of continents, the chief areas being in southern China, the Gulf-Atlantic States of the United States of America, and in South-eastern Australia. There is well-distributed rainfall throughout the year, and the winter months have mean temperatures above 43° F., so that the forests are evergreen and almost as luxuriant as the tropical rain forests. The trees include the evergreen oak, mulberry, maple, walnut, and various eucalypts; in some areas lianas are common (Fig. 231). On poor soils conifers are also found in these forests.

3. Cool Temperate and Cold Forests

Amongst these are the deciduous hardwood forests and the coniferous forests, but there are extensive areas of mixed forests. Both deciduous and coniferous trees are compelled by a cold season to rest in growth for a period of several months which increases in length in a polewards direction.

(a) *Deciduous Hardwood Forests.*—These are found mainly south of the coniferous forests in the cool temperate regions and on the margins of the two great northern land masses. The period of cold, when the forests are leafless, is usually less than six months. The trees require time in spring for the development of fresh leaves before full

---

[1] Eucalypt means well covered, in that the flowers are protected by a cap.

growth is possible. The leaves are broad and the trees can transpire freely during the growing season, as the summers are quite moist. The forests have much undergrowth in parts. The oceanic western margins are not so suitable for deciduous trees as the sub-marginal areas, since strong winds on the coasts promote too rapid transpiration and salt spray is inimical to the growth of trees. These forests persist eastwards in Europe, but decreasing summer rainfall causes them to fail in western Russia. In both North America and Asia simi-

Dorien Leigh Ltd.

FIG. 204.—Virgin Deciduous Forest in Germany
Beech in the foreground and oaks behind.

lar forests are found on the eastern margins. These forests include trees such as the oak, beech, and ash, as well as the maple, chestnut, and elm (Fig. 204).

(b) *Coniferous Forests.*—These are found mainly in two great continental belts across North America and Eurasia, in latitudes which reach northwards to the tree limit that is determined by low temperatures, so that there is too short a period for their growth (apart from dwarf trees in the tundra). In general, the coniferous trees have a growing period, with mean temperatures above 43° F. of not more than six months. This period is too short for deciduous trees, which require time for leaf formation, but the needle leaves of the conifers remain during the winter, and with the onset of temperatures above 43° F. they are able to start growth immediately. The summer rainfall of the interiors is sufficient for conifers but not for most deciduous trees, although a few of these persist into the far north. The chief types of tree are pine, fir, spruce, hemlock, and various larches. On the western mountains of North America in middle latitudes occur forests

Dorien Leigh Ltd.

FIG. 205.—Redwoods in California, United States of America

Note the fairly sparse undergrowth of small trees.

with very large conifers, including Douglas fir and, in California and Oregon, two trees, the redwood and the big tree, both belonging to the Sequoia family (Fig. 205). The latter trees are probably the tallest in the world, only being rivalled by the jarrah and karri of Australia. Other conifers, such as cedars and cypresses, are not uncommon in the Mediterranean areas and in isolated highlands in the tropics, such as in Abyssinia and East Africa.

In coniferous forests there is little undergrowth as compared with the deciduous hardwood forests which have many shrubs between the trees. The evergreen nature of the conifers and the low sun angles in high latitudes during most months help to restrict the sunlight from reaching the ground so that undergrowth is discouraged.

It should be noted that the deciduous hardwood forests usually merge with mixed forests of deciduous and coniferous trees on the poleward margins. In the coastal plains of the Gulf-Atlantic States of America, southern long-leaved pine forests are found on sandy soils adjacent to the broad-leaved evergreen forests mentioned above. In general, broad-leaved forests call for better soils than do coniferous forests; consequently there may be found areas of conifers on pervious and infertile soils within deciduous forests.

Grassland

This type of vegetation is composed essentially of perennial grasses in tufts which form a fairly continuous cover to the land surface. Grasses are herbs and may have deep roots, but they have no woody structures as do trees. Other herbaceous plants are frequently found in grasslands. Grasses are usually shallow-rooted plants, and during the growing season they must receive frequent, rather than heavy, rainfall to replenish moisture which is lost by evaporation from the surface layers of the soil. The matting of most grasses is helpful in retaining moisture near the surface. Grasses have a rest period in their growth, during which it may be said that the prevailing climatic conditions are of no importance whatsoever; they simply wither and die down until the return of more humid conditions when they flourish again. Drought in the growing season is detrimental to most grasses. One exception is Mitchell grass in Australia which is very drought-resistant.

The most suitable areas for the development of natural grasslands are the continental interiors, since these rarely have drought in summer, the vegetative period for grass. Grasslands show a gradation in their luxuriance which is roughly proportional to the amount of rainfall received in the growing season. With decreasing rainfall grasslands thin out, becoming steppe which is not uncommon in semi-desert regions. 'Steppe' is used to describe both tropical and temperate grasslands which have scanty rainfall.

Grasslands may be grouped in the following manner:

1. *Tropical Grasslands.*—These are typical of the tropical continental interiors where the rain is concentrated in four or five months of the hot season. With abundant rainfall at this time, grasses may grow to a height of twelve feet or more. There is a considerable range of vegetation in these grasslands and trees are quite common. In fact,

Black Star

FIG. 206.—Tropical Grassland in Australia

savana, which means a treeless plain, is often confused with areas that are more suitably described as grass woodlands (see p. 279). These tropical grasslands pass equatorwards, with increasing rainfall, into dry forests, and polewards, with a longer season of drought into steppe and semi-desert (Fig. 206).

2. *Temperate Grasslands.*—These are found in the continental interiors in both warm and cool temperate latitudes with spring and summer rainfall. They merge almost imperceptibly with the sparse grasses of the steppes of the cool or temperate deserts. The richer grasslands of these regions are known as prairies, and contrast with the steppes which have shorter grasses, although in both types of

grassland there are tall and short grass formations. As in tropical grasslands, these are accompanied by trees. On the eastern margins of the Canadian prairies isolated oaks, elms, and maples persist from the eastern forests, together with groves of poplars and willows. However, with decreasing rainfall to the west the trees fail and a true grassland formation is established with some stunted shrubs and cactuses. Prairie grassland is shown in Fig. 207.

3. *Meadow Grasslands.*—These are common in the temperate margins which have neither a season of drought nor severe winter

E.N.A.

FIG. 207.—Prairie Grassland, Sierras de Cordoba, Argentine

cold. They have developed with the clearance of forests in cool temperate areas in particular.

Desert Vegetation

Under desert conditions, aridity restricts the vegetation to such an extent that a complete plant cover is absent. Whilst in some desert areas plant growth is well-nigh impossible because of extreme aridity, most deserts have a representative flora of their own. The areas of waterless desert are very limited, and much of the regions shown as desert on maps is better defined as semi-desert from the point of view of the vegetation. One characteristic of this vegetation is the isolation of the individual plants, each being separated from its neighbours by considerable distances (Fig. 208).

The vegetation is truly xerophilous, and all plants in the deserts and semi-deserts must seize upon every opportunity to further their existence whilst being adapted to resist or endure the effects of prolonged drought which may last for years rather than for months. The seeds of some desert plants are capable of remaining inert for years until the chance incidence of rainfall gives them the opportunity of a brief life. Grasses are not uncommon, but they have specific adapta-

Black Star

FIG. 208.—Desert with very Sparse Grass, Negev, Palestine

tions for the purpose of drought resistance, such as thickened outer layers, inrolling leaves, and the development of many hairs.

There is a tendency for the plants of the marginal zones, just outside the deserts and semi-deserts, to persist within the areas. For instance, the drought-resistant trees of the acacia family range from the tropical grass woodlands into the semi-deserts. In Australia both the mulga and the brigalow are acacia thickets which lie within the semi-desert regions. The cactus is usually regarded as typical of deserts. It is, by origin, an American plant, but has been successfully introduced into other continents and into regions with more rainfall than is usual in deserts. With its leathery surface, covered with wax and resin, and with its prickles, which are actually its leaves, it reduces transpiration to a minimum, and as a succulent it has distended cells which store much water.

Many plants, including the cactus, have most extensive root systems which may extend for many yards underground, and thus, although the parts of the plants which are visible above the ground are quite restricted, each plant, taken as a whole, cannot be regarded as stunted. By means of their long roots it is possible for many of these plants to collect very large quantities of water to be stored in various ways. Sand has often very great reserves of water at depth despite its very dry surface layers; some of this water is derived from areas of greater rainfall, and has travelled underground to the desert regions.

Probably the most difficult task of desert plants is to ensure that their seeds are scattered and germinated. Some seed pods are preserved in a round cage, made of dried branches, which uncurls only upon contact with moisture. This small cage, or ball, is blown across the desert, perhaps for years, until an opportunity for germination presents itself. Some plants produce succulent, water-laden fruits, the seeds of which may be dispersed when the fruit is eaten by animals.

The desert vegetation of the world may be divided into two main groups, belonging to:

1. The hot deserts and semi-deserts which lie in the arid trade-wind belts of the tropics and sub-tropics.

2. The cool or temperate deserts which are located in the intermontane basins in the western parts of North America and in Inner Asia. The mountains surrounding these basins isolate them effectively from rain-bearing winds.

The hot deserts and semi-deserts may be subdivided further into those with rain at one season of the year and those which have at least two seasons of rain in each year. The first sub-group has no great variety of plants, and does not include succulents, which are found in the wetter deserts, although even the latter have relatively few types of plant (Fig. 209). The plants of these regions may be divided into four categories:

(a) Drought-evading plants, which are annuals whose seeds await the onset of fresh rainfall. They include desert plantains and desert fescue.

(b) Drought-resistant plants, such as acacias and the mesquite bush of North America, which have long roots that are able to reach a deep water table and therefore flourish in an apparently arid zone.

(c) Succulents which have a water store within their tissues and shallow roots that spread extensively in a lateral direction in order to provide as large a collecting area as possible when rain falls. They are seen to swell with an income of fresh moisture. They have a very slow growth, but they are often long-lived; even after uprooting a cactus may live for several years, drawing almost imperceptibly on its water store. The cactus and the euphorbia are the commonest of the succulents.

(d) Drought-enduring plants, such as the creosote bush of the North American deserts, which have no deep roots, no water storage,

Black Star

FIG. 209.—Desert, Arizona, United States of America

The Segaure cactus and other plants give a fairly rich desert vegetation. On the hills in the background is fairly typical chaparral.

and do not evade the drought as do the annuals. They appear to possess a unique ability to recover rapidly after wilting, with a slight increase in the relative humidity of the air such as occurs after sundown, or with the onset of rain. This remarkable behaviour is due to a significant and unusual character of the protoplasm of the plant. If this character could be transferred to cereals, it is possible that man might extend the areas of cereal cultivation into quite arid regions.

In coastal areas on the west of the hot deserts some plants exude hygroscopic salts from the leaves, which permit absorption of moisture from sea mists and fogs.

Tundra

Finally, a few remarks must be made about tundra, which lies in the far north of the world. Tundra has often been described as a form of desert, in that the vegetation suffers from a physiological drought because low temperatures prevent the plants from drawing on mois-

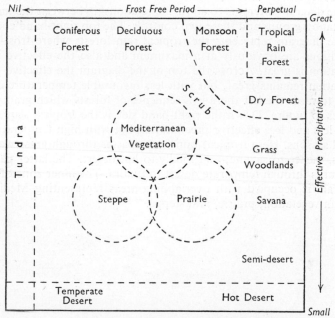

FIG. 210.—A Summary of Vegetation
The author is indebted to Professor E. Ashby for this diagram.

ture which is in the solid form of ice. However, modern investigations cast doubt upon the reality of a physiological drought, and it is suggested that the plants of the tundra are adapted to the peculiar climatic conditions of the regions. The vegetation consists of some grasses, but mainly of mosses and lichens. Most plants, including some dwarf trees, are xerophilous in appearance although not in character (Fig. 248).

Further details of vegetation will be given in the section of the book which deals with the climates of the world (see Chapters XVII–XX).

A Summary of Vegetation

To conclude this chapter, it may help the reader to study the diagram (Fig. 210) above. It takes account of two significant controls on plant growth:

1. The length of the frost-free period in the year.

2. The effective precipitation, which may be regarded as the moisture available to the soil and plants after allowance has been made for loss to both through evaporation and run-off.

The diagram attempts to show the relationship between the major vegetation groups and these two controls. The top right-hand corner of the diagram represents the tropical rain forests where frost-free conditions are obviously at a maximum and also the effective precipitation is great. Across the top of the diagram the effective precipitation remains great, but with less favourable temperature controls towards the left there is a sequence of forests which gradually passes to tundra. Down the right-hand side is the tropical sequence with less and less effective precipitation but with high temperatures at all seasons. Here tropical rain forest passes through grass woodlands, thence to savana, and finally to hot desert. The latter passes leftwards through temperate desert to tundra. The inner zone of the diagram is occupied with overlapping areas representing Mediterranean vegetation, prairie, and steppe.

CHAPTER XVI

SOILS AND SOIL EROSION

Soil

The superficial layers of the earth's surface usually consist of soil which is a mantle covering the solid rock below. Soil represents the product of prolonged weathering, and the depth of the soil is fairly closely related to the time during which this process has gone on. Initially, the soil consists of fine mineral particles which are developed from the rock surface by weathering as described on pp. 6 and 7. Once a layer of such fragments has been produced, it is possible for plant life to establish itself on the surface of the soil, and within the pore spaces of the soil bacteria find a suitable environment in which to live. On the death of plants their tissues decay, and this process is materially assisted by bacterial action, which creates various compounds so that a supply of organic substances is added to the minerals of the soil and thus it is enriched. Many of these products as well as some of the mineral substances are capable of being dissolved by percolating rain-water and they are carried downwards in solution to affect the lower layers and to break up still more of the underlying rock by chemical action. The deepening of the soil is clearly dependent on an abundant supply of water, although this alone is not sufficient to develop a good soil, but it is in regions of copious rainfall that the most extensive soil layers are likely to develop. The very deep tropical soils of 40 feet thickness owe much to the abundant rainfall combined with the high temperatures of these regions.

Mineral Particles of the Soil

The mineral particles which are found in the soil are produced mechanically by the action of heat, frost, and wind. Some of these particles may be further changed by circulating rain-water which introduces the chemical processes of oxidation, reduction, hydration, and carbonation. It was shown in Chapter I that the essential end-products of such weathering of granite were clay and sand particles, with carbonates in solution. The size (diameter) of the fine particles in a well-developed soil ranges from 2 mm. down to less than 0·005 mm. The coarsest of these are known as gravel and sand, those with diameters between 0·05 and 0·005 mm. are known as silt and the finest which are less than 0·005 mm. in diameter as clay. It should be noted that these names are used to define the size of particles, and that they do not necessarily indicate the mineralogical nature of the particles.

It will be seen that the nature of a soil must be related to the geological character of the underlying rock or the transported rock fragments which may cover the bed rock. Whilst the rock underneath plays its part, it is far from correct to assume that the final or mature soil is determined mainly by geological factors. Other constituents and various processes, to be described, help to produce much the same soil over large areas irrespective of the various rocks below, although the latter may play a predominant rôle in soil formation in some localities.

Organic Particles of the Soil

It has been noted that the upper layers of the soil are a source of supply of organic compounds. On the death of a plant, profound changes are wrought in its material, chiefly by the action of bacteria and, in certain circumstances, by fungi. These changes are mainly directed to the breaking down of the plant refuse into somewhat simpler substances, and then the development of new and often quite complex substances. The remains of dead animals, such as those of worms and other burrowing creatures, will also undergo similar changes. Resulting from these various changes is an organic product known as humus, a complex colloidal or jelly-like material which is brown in colour. The presence of humus in the soil permits yet further changes to occur. It must be realized that a plant cannot utilize substances unless they are in solution, and humus has the power of assisting in the formation of such solutions that are so vital to the plant. Furthermore, humus encourages bacteria to flourish, since it is a food for these micro-organisms. This property of humus must obviously lead to an increase in the number of bacteria which are likely to reside in the soil, and so there is a further increase in organic material.

Bacterial Action in Soil

Bacteria of certain kinds find a lodgment on the roots of plants such as clover, beans, and legumes. Certain characteristic nodules on the rootlets of these plants are indicative of bacterial colonies. The prime function of these particular bacteria is the fixation of nitrogen from the air which permeates the soil spaces. Nitrogen is an important element in plant nourishment, but plants can only use it in solutions of nitrates and ammonium compounds, the latter being converted to nitrates and nitrites.

Bacteria are reduced in numbers in the soil if the air supply is poor, consequently humus is not readily developed. A soil which is poor in oxygen, due in all probability to water-logging, is known as anaerobic, and, with a drastic reduction of bacteria, plant refuse in such a soil is hardly changed at all, but remains as a layer of material which has a markedly acid reaction. The very important compounds, derived from soil particles and by fixation of nitrogen, which give the essential solutions on which a plant is to feed are substances containing potassium, nitrogen, and phosphorus, with some calcium and sodium, in conjunction with oxygen. It is known that there are required for successful plant growth additional compounds containing what are known as trace elements, such as iron, manganese, cobalt, and boron, but the origin and function of these elements is still obscure.

Characteristics of a Good Soil

Since the soil is the medium in which plants grow, it is of value to assess the qualities which make a good soil. These may be summarized as follows:

1. In order that plants may feed, solutions of various compounds are required, and this of necessity calls for an adequate supply of soil moisture.

2. For the provision of humus and other organic products air, with its oxygen for the support of bacterial life and with its nitrogen to provide by bacterial activity nitrogenous compounds, must permeate the soil.

3. Soil must therefore have an abundance of spaces to permit the free circulation of both moisture and air. This circulation is most effective when the soil possesses a suitable texture, which is largely determined by the size of the mineral particles, for these may be regarded as nuclei around which various compounds collect in solution. Furthermore, the soil may have a flocculated structure which is produced by the grouping of individual particles into large groups or floccules. Such is known as crumb structure.

Soil Profiles and Leaching

It is customary to describe a soil by reference to its profile. This is demonstrated by a vertical section through the soil to the underlying bed-rock. In most cases such a section will show horizontal layers of varying thicknesses, which are often distinguished from one another by their different colours. Each distinct layer is called a horizon of the soil, and the horizons usually form two groups, the A group at the top and the B group below (Fig. 211). The underlying rock is often referred to as the C horizon (Fig. 212). In general, water in the soil moves downwards, and the solutions of organic compounds and mineral salts are frequently transferred from the A horizons to the B horizons. In regions of abundant and heavy rainfall, this is a process of great significance, and results in the leaching of the upper horizons. This means that there is eluviation (washing out) of humus and minerals or a depletion of much plant food in these upper horizons. Clearly the lower horizons, in particular the B horizons, are enriched by the illuviation (washing in) of these solutions and minerals. Such

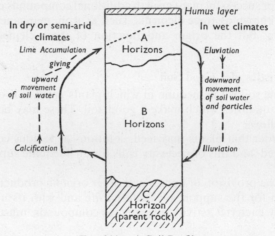

FIG. 211.—A Soil Profile

No scale is given as relative thicknesses of horizons vary with profiles.

soil profiles show a darker-coloured B horizon as compared with the A horizons, which are lighter in colour as a result of this depletion. Usually the top of the A horizon is fairly dark, because there is much humus near the surface, and this is not all leached by the rain.

Podzolization and Laterization

In cool and cold lands the leaching process is known as podzolization and the removal of the humus and iron oxides from the upper layers leaves little there save silicates and oxides of aluminium. This produces a characteristically pale, sometimes ash-grey, coloured surface horizon which is typical of the podzol of the Russian coniferous forest zone from which the term is derived. Under tropical heat the

Geological Survey photograph reproduced with the permission of the Controller of H.M.Stationery Office

FIG. 212.—Quarry, near Tintagel, Cornwall

Soil developed above a parent rock, in part, of lava.

process of leaching is on a grand scale and is called laterization. The resultant soil is red or yellow in colour near the surface, although sometimes leaching in the tropics produces a whitish surface layer which is essentially bauxite. This material may be regarded as a rock, rather than a soil, and is very rich in aluminium oxides. However, it is rare for the iron oxide to be completely eluviated from the A horizons, and thus the more usual reddish appearance of the tropical soils is explained. Quite obviously there is an accumulation of aluminium and iron oxides in the B horizons of such strongly leached soils, and yet, despite this depletion of the A horizons, there remains a considerable residue of aluminium and iron compounds, so that such soils may be called pedalfers, i.e. soils with aluminium (Al) and iron (Fe).

Calcification

It is possible for water in the soil to move upwards. This occurs in areas of desert or regions with a small rainfall. The upward lift of the water is effected by capillarity. Water which has sunk by gravity may be drawn up through the pore spaces of the dry upper horizons. This process is comparable to the absorption of water by a good cloth as used in drying crockery. The water on the damp crockery is not wiped off so much as drawn into the fine pore spaces of the cloth. Such water, in the lower layers of the soil, is often heavily charged with dissolved carbonates and other salts, and upon reaching higher levels it usually evaporates and leaves an accumulation in the A horizons of solid calcium carbonate and other salts. The drier the region in which this takes place the higher is the zone of accumulation, and in some areas of desert the surface may be encrusted with salts derived from below. This process is known as calcification (Fig. 211).

The soils which show by upward movement of the soil solutions an accumulation of calcium carbonate, that may outweigh in importance the other minerals of the higher horizons, are known as pedocals, i.e. soils with a calcium (Cal) content. In regions of low rainfall it may be seen that the level of the deposits of calcium carbonate varies with the rainfall and usually the highest levels occur where the rainfall is least.

Acidity and Alkalinity of Soils

Carbon-dioxide is dissolved in soil water, some of it being extracted from the air and some produced by bacterial action. The dissolved gas forms carbonic acid (H_2CO_3). This slightly acid solution may become more markedly acid in cases where organic acids are formed also by humus in the soil. On the other hand, there are usually compounds of calcium and sodium in the soil which have an alkaline reaction. Soils may be classified according to their pH value. This pH value expresses the concentration of hydrogen ions in the soil solution. With a neutral solution the value is 7 when the concentration is, very nearly, 10^{-7} grams of hydrogen ions per litre of solution. This figure 10^{-7} may be regarded as a fraction, $\frac{1}{10^7}$. An increase in this concentration of hydrogen ions indicates acidity. For instance, with a pH value of 6 the fraction is $\frac{1}{10^6}$ which is ten times greater than $\frac{1}{10^7}$. On the other hand, a pH value of 8 means a smaller concentration of hydrogen ions. Acidity in the soil is therefore expressed by values of pH which are below 7 but usually not less than 4·5. When the hydrogen ion concentration decreases so that the pH value exceeds 7, the soil is alkaline. In general, soils range from 4·5 to 9 as regards pH values. With most crops it is found that slightly acid soils with pH values of 6·0–6·5 are best. Too acid a soil produces unfavourable conditions; in particular, bacterial action is restricted. On the other hand, alkaline soils do not permit plants to avail themselves of certain trace elements.

Liming of the soil may be necessary to reduce soil acidity to some extent. It is quite erroneous to regard liming as effective only when it

creates strong alkalinity, because such a state is as undesirable as considerable acidity.[1] The calcium carbonate content is high in a pedocal
because of the calcification, and the surface horizons have pH
values well above 7, and this value may be increased through greater
concentration of certain calcium and sodium compounds when the
soil becomes literally saline.

Textural and Chemical Classification of Soils

It is possible to classify soils in several ways. To a farmer it is not
necessarily important that he can distinguish a pedocal from a pedal-
fer, nor is it imperative that he knows that a soil has been podzolized
(or that it has undergone laterization if he is within the tropics). The
farmer is much more concerned with the texture and chemical reaction of his soils; in fact, it may be argued that an adequate knowledge of either of these soil characteristics is sufficient for practical
purposes. On the other hand, more detailed classifications must have
reference to the soil profiles. World-wide soil classification is based
on the profile of a soil, and results in the recognition of broad zones
of one soil type which may be continental in extent.

The interest of the farmer in the texture of his soils has led to a
textural classification. On p. 255 reference was made to the dimensions of the mineral particles in soils, and three types of particles
were noted, namely, sand, silt, and clay. The texture of a soil will
depend primarily on the proportions of the various particles which
are included within it. It is possible to define a variety of intermediate
types of soil with textures lying between the three main types, sand,
silt, and clay, as shown by the diagram (Fig. 213). The corners of this
diagram represent soils with 100 per cent. clay, with 100 per cent.
silt, and with 100 per cent. sand particles respectively. Nearly equal
proportions of sand and silt, together with a little clay, produce a
loam soil which passes on the one hand to sandy loam and on the
other to silty loam, with less silt and sand respectively. Similarly, sand
and sandy loam develop into sandy clay loam and sandy clay, and
finally clay, when more than 50 per cent. of the particles are of clay
texture. Silt, with increasing clay particles, passes to silty clay loam,
then to silty clay, and finally to clay and so on.

The texture of the soil is important, for soils with fine particles have
larger particle surface areas than soils with very coarse particles, and
this affects directly the economy of the plants in regard to the available food, since a larger surface area means a greater food supply in
the form of aqueous solutions. A certain structure is developed by
the grouping of the particles into masses. This flocculation, as it is

[1] Liming should normally produce a pH value between 6·0 and 6·5.

called, is very desirable, because it ensures the permeability of the soil, and thus keeps it effectively aerated. Calcium carbonate, humus, and other organic products are of value in creating this flocculation. Some clays are impermeable unless they have been artificially flocculated by the addition of lime or organic fertilizers.

The classification of soils by their chemical reaction depends upon the determination of the pH values of the particular soils. Much work has been done in correlating the success of various crops with the pH

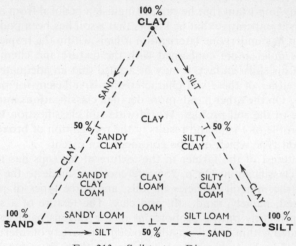

FIG. 213.—Soil-texture Diagram

values of the soils in which they flourish, and undoubtedly such relationships are invaluable to the farmer, but it is outside the province of the geographer to go further into this study.

Geographical Classification of Soils

When discussing the world distribution of soils, it is customary to refer to the major soils as zonal, in that very often they cover great tracts of continental surfaces, and their characteristics are largely dependent upon climatic controls. The parent rock has little influence on the type of zonal soil which is found in a given region. Thus, a soil derived from a granite parent in the tropics is unlikely to resemble a soil, also with a granite parent, which is developed under cool or cold climates. When it is remembered that the end-products of the disintegration of rocks are essentially particles of sand, clay, and carbonates (see p. 7) and that any rock will yield some, if not all, types of mineral particles for soil production this may seem rather extraordinary. However, it is the predominance of some particular process, such as leaching or calcification, which is primarily responsible for the type of zonal soil which is produced, and it is therefore the climatic control that is of paramount importance. Under wet climates pedalfers are common as a result of podzolization or laterization, whereas in arid or semi-arid regions pedocals will be typical.

In addition, there are intrazonal soils which may cover extensive areas within the regions of various zonal soils. They depend usually upon some special factor or process. There are three groups to be mentioned:

1. Soils in which salinization has occurred. In this case, as with calcification, a dry climate results in a concentration of soluble salts in the soil. Soils which contain a very high proportion of salts are known as solonchaks. However, with moderate leaching a solonchak may be converted into solonetz, which is a relatively alkaline soil on account of included sodium hydroxide. The solonetz may in its turn be effectively leached, so that a soloth is produced which shows some degree of podzolization. The colour of these various soils is usually grey or white and often little structure is visible.

2. Soils in which gleying or gleization has occurred. In this case water-logging, near rivers and in bogs or in the swamps of forests, creates these special soils. They are frequently dark at the surface owing to a high organic content but below they are usually blue and green in colour, a result of water-logging which excludes oxygen, thus producing chemical reduction.

3. Rendzinas (terra rossa) which are developed on limestone or chalk rock. Usually these soils are very shallow and vary in colour from dark grey to yellow and reddish brown. Terra rossa is a term used widely for any soil of reddish hue.

Finally, there are azonal soils which are of very slight importance. They may be considered as barely changed rock. Glacial ground moraine, screes at scarp feet and at the base of mountain slopes, water-logged river alluvium of recent formation, and dry sand of dunelands are of this type. They may be regarded as skeletal soils since they often possess the necessary mineral content for the formation of good soil, but lack the organic fraction which is so necessary. Loess is technically a rock, and not a soil, but it is possible to have from it a development of loessic soil on account of the richness of loess in minerals and soluble salts. High mountain zones are somewhat exceptional in regard to their soils (as they are from a climatic point of view) and many soils of such regions are skeletal in form. Steepness of slopes and extremes of temperature have their effects on the processes of soil development as well.

The outstanding soils of the world are discussed in relation to the climatic regions in which they occur (see pp. 294, 309, and 329).

Soil Erosion

Soil erosion is an evil which has affected many areas of the world in this century, and whilst it is largely man-made, it can be legitimately included in a study of physical geography. Under a cover of natural vegetation the chances of destructive soil erosion are practically nil. Apart from desert areas, removal of large quantities of surface rock waste in a relatively short time is almost negligible. There has been, over thousands of years, a vast removal of rock waste and earlier developed soils as rivers have done their work, but it has to be noted that this process has been extraordinarily slow. But, in a matter of a few decades, man has destroyed the work of nature which has taken so very long.

Once man has removed by cropping or over-grazing the natural mat of vegetation, particularly in grasslands where the lack of trees enable wind erosion to be so effective, then soil erosion has been an almost certain consequence. Thus dust bowls are literally man-made deserts where deflation has proceeded very rapidly. However much soil erosion has been caused by rain storms and this has been most common in semi-arid regions. The continental grasslands in temperate and, to a large extent, in tropical latitudes have suffered most of all where they have been cropped or where the grass has been thinned by heavy grazing. The soil erosion in these regions is certainly aggravated in part by the incidence of the rainfall in short but very heavy showers, which create a considerable run-off once the surface layers of the soil have become temporarily water-logged.

Apart from wind erosion there are two main types of soil erosion which can be regarded as truly destructive. Firstly, there is sheet erosion, in which a uniform layer or depth of the soil is removed by rain washing down a slope. This is much encouraged if cultivation has led to ploughing up and down the slopes, because the individual furrows become shallow gullies along which the soil moves with the water. At the foot of such slopes there may accumulate soil deposits which are stratified as a result of successive rains. This is a very widespread form of soil erosion, and it is not easily detected in its earliest stages because the surface of the land shows no radical alteration in appearance. The degree of sheet erosion depends upon several factors, such as the angle of the slope, the texture of the soil, and local climatic conditions, together with the methods of tillage which have been employed.

The second type of erosion is known as gully erosion, and whilst wrong methods of cultivation may lead to it, more usually it is local relief which sets up this type of erosion by the establishment of a small-scale drainage system that is regularly used by each additional rush of water after a heavy storm of rain. The effect of gully erosion is to create 'bad land' surfaces, and each gully extends in a headward direction with later erosion, so that a dendritic pattern is produced. Such gullying may be seen very frequently on spoil heaps in industrial regions of this country. The waste of rock and soil is carried from the land down these gullies, and collects as a fan of soil and stones which may unfortunately overwhelm and destroy the value of other adjacent crop lands that have not so far been eroded.

Soil Conservation

Various methods have been adopted to check soil erosion where it has commenced, and these are mainly devices to prevent the downward movement of the soil on slopes. It is almost impossible, with cropping of the land, to avoid having the bare soil exposed during certain periods of the year, but one preventive method is known as strip cropping, in which the areas of the soil exposed at any time are restricted. They are separated by strips carrying a vegetative cover of grasses or some leguminous crop. The down wash of the soil is also prevented by contour ploughing and terracing. In the first method the furrows are ploughed by following the contours of the land, which means that the furrows are rarely straight but in sweeping curves. Each ridge between the furrows is therefore level and lies across the down slope. The second method is terracing, which is devised to fulfil one of two purposes. Either the terraces trap the water so that it is gradually absorbed into the soil, or they deflect water, which would otherwise wash down-slope, into drainage channels that carry it away. In the former case each terrace is level and has soil piled up to retain and absorb the water, and in effect this is the same as contour ploughing. In the latter case the terrace has a low ridge of soil on the down-slope side, and the terrace slopes laterally across the down slope into a drainage channel. Another method of preventing soil erosion in crop land is to use a lister with a basin-forming attachment. The lister is primarily for setting seed. Its front part produces a broad furrow, but it is further improved by the addition of the basin-forming attachment. This creates a series of basins in each furrow by throwing soil athwart the line of the furrow. These soil dams hold the rain in small basins or pools which gradually drain into the ground. Where wind is a predominant agent of erosion, the planting of wind-breaks is a measure to restrict erosion. Flood control of large rivers is closely linked with these various efforts to prevent or arrest erosion.

Part V

CLIMATOLOGY

CHAPTER XVII

INTRODUCTION—HOT CLIMATES

Climatic Classification

Attempts at a classification of climate may be said to date from the time of the Greeks, who had a qualitative scheme involving thermal belts, but it was not until the late nineteenth century and even later, in this century, that quantitative classifications of climate were formulated. Supan in 1879 attempted to correlate the boundaries of the Greek thermal belts with isotherms which had appeared on a map, for the first time, in 1848. Later, in 1884, Köppen, whose name is inevitably linked with climatic classifications, modified these belts of Supan.

However, Köppen was concerned, primarily as a botanist, in an attempt to establish fairly rigid limits to the major vegetation zones of the world, using temperature and rainfall values. This led him to publish his classifications of 1900 and 1918. In the former some of the climatic types which he recognized were actually named after plants, e.g. baobab, liana, and birch. In trying to fit the salient vegetation types within his limits of temperature and rainfall, Köppen may be said to have been too rigid. Consequently we must accept the evidence of later workers, such as Russell and Thornthwaite in America, who have shown that clear-cut climatic boundaries are very rare, and that just as one distinct vegetation zone merges into another through a transitional zone, so climatic types must be allowed such transitions. It is quite possible that, within a period of a quarter of a century, the controls which are deemed to separate two distinct climatic types may change, and it is therefore reasonable to assume a shift of the accepted boundaries in such a period. However, there is no doubt that the major vegetation zones as well as the major soil regions reflect very closely the control of climate.

In this book, limits of temperature and régimes of rainfall are used to determine the climatic regions which are closely related to the natural vegetation regions and, very largely, to the regions of zonal soils. It has been considered of value to discuss both vegetation and soils separately in previous chapters, but in the following section of this book further discussions of both topics are given in appropriate places.

The mean monthly temperatures and the mean monthly rainfall totals may be used to show graphically the essentials of the climate of a given place. Throughout the following chapters such graphs are employed to illustrate the text, and it may be mentioned that all the graphs are on the same scale as shown in Fig. 214, and that all carry the same ancillary information, namely, the mean annual range of temperature and the total annual rainfall as well as the height of the place above sea-level. The graphs are placed beside the several maps

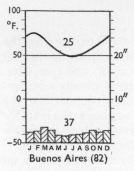

FIG. 214.—Climatic Graph

which attempt to show the regional distribution of the climatic types and sub-types that are reviewed. Where places are referred to in the text, it will be found that, in most cases, there is the climatic graph for that particular station in the same chapter.

Also, the maps show the related natural vegetation, but in some parts of these maps it will be seen that overlaps are indicated by the shading employed for a given type of vegetation. It has been observed above that transitional zones do exist, and that it is well-nigh impossible to develop a scheme of climatic boundaries which perfectly fits all vegetation zones. The purpose of the various maps is to show as far as possible the vegetation which lies within the limits of temperature and rainfall that have been adopted.

The classification which is adopted is as follows:

1. Hot climates—with mean annual temperatures (reduced) of at least 70° F.

2. Warm temperate climates—with the coldest month having a mean temperature of not less than 43° F.

3. Cool temperate climates—with one to five months having mean temperatures below 43° F.

4. Cold climates—lying polewards from the regions with temperate climates and divided into cold temperate, tundra, and polar climates by various temperature controls, as will be seen in due course.

Several subdivisions of each of the above types, excepting the tundra and polar climates (since no sensible distinctions can be made), are determined by the régime of rainfall and, in some cases, by the occurrence of monsoonal controls.

Hot Climates

(a) Introduction

The areas with hot climates are enclosed by the mean annual isotherm of 70° F. (see Fig. 216). Most of this zone is tropical, but parts are extra-tropical and extend to approximately 30° N. and S. The areas are thus within the trade-wind belts of each hemisphere, and include the Doldrums belt with the intertropical front. Three major types of climate with their subdivisions may be distinguished as follows:

1. *Equatorial.* With rain at all seasons, and sometimes exhibiting a double maximum of rainfall.
2. *Tropical.*   (i) Continental, with rain in the hot season only.
        (ii) Marine, with rain at all seasons, but with a distinct maximum of rainfall at one period of the year.
     (iii) Monsoonal, with rain mainly in the hot season.
3. *Hot Desert.* (i) Continental, with little or no rain.
      (ii) Marine (west coast), with little or no rain, but much influenced by cold currents offshore. In part, areas with this type of climate lie outside the hot regions as defined above.

(b) Equatorial Climates

Main Controls

Regions with these climates are quite limited in extent, rarely stretching through more than ten degrees of latitude in all. Two characteristics of the climates are the constant high temperatures and the well-distributed rainfall, so that, in general, there is no season of drought. The control of the Doldrums belt with its constant temperature, low-pressure air, and regular convectional rainfall has long been regarded of primary importance. However, air masses, whose sources are in the subtropical high-pressure cells, converge along the intertropical front, where air ascends in keeping with the earlier conception of convectional uplift in these areas. The humid warm air masses of oceanic origin are unstable, and productive of very heavy rainfall. The intertropical front oscillates north and south with the annual movement of the overhead sun, although it coincides fairly closely with the heat equator. In the northern summer it is well north of the geographical equator, so that there are marked transgressions by the south-east trade winds into the northern hemisphere, and both the Guinea coast of Africa and parts of the East Indies are influenced by air streams which have their origins in the southern hemisphere.

Temperature

Insolation is fairly uniform throughout the year, since the noonday angles of the sun are never less than 66½° (at the solstices). Mean monthly temperatures remain very steady in lowland areas at about 80° F., whilst in neighbouring uplands the values are about 55°–60° F. (see Quito).[1] Mean annual ranges are remarkably small, rarely more than 5° F., and in oceanic positions they may be less than 1° F. Absolute maximum and minimum temperatures are of the order of 100° F. and 60° F., respectively, in lowland regions. The keynote of the temperature is its monotony both as regards monthly and daily rhythms. Two equinoctial peaks of temperature may coincide with overhead sun (see Georgetown), but this feature is actually far from common.

Rainfall

Whilst the Doldrums belt is a zone of convergence, there is little contrast between the air masses, and thus frontal developments are unusual, apart from occasional squalls. The fairly constant insolation produces a regular convectional uplift of air, but rain does not necessarily fall every day and some months may have little rain. Rain usually falls between noon and 4 p.m., and follows cloud formation which is consequent upon the mounting effect of convectional heating during the earlier hours of the day. Annual totals of rainfall are commonly between 65 and 80 inches, but local relief may produce greater totals, and peculiarities of régime are shown by many stations.

It would be expected that a maximum of rainfall should appear at each equinox or shortly afterwards, when convectional uplift should be greatest in view of the passage of the overhead sun, but there is usually a lag between zenithal sun and the highest temperatures. Such a double maximum is actually not very common, because sometimes one peak of rainfall is entirely suppressed (see Para) or is much smaller than the other.

With the movement of the overhead sun between the tropics there should be theoretically two periods in each year (about the solstitial dates, when the Doldrums rain-belt is shifted farthest north and south) when the rainfall should be less than at other times in equatorial regions. In fact, just north and south of the equatorial zone with well-distributed rainfall the régime shows two drier seasons (not rainless seasons) when the sun is overhead at either tropic (see Yaundé). Still farther polewards the drier seasons are more marked, but they are unequal in length, and still nearer each tropic the interval between the two passages of overhead sun becomes shorter, and by about 10°–15° N. the two rainy periods coalesce and there are, in essence, the two seasons in the year, one dry and the other rainy, of the tropical continental climates.

[1] Climatic graphs are on pp. 276-7.

However, the Doldrums belt and the intertropical front are not placed symmetrically about the geographical equator at the equinoxes as would be expected. The heat equator is displaced northwards for most of the year, particularly in Africa, and to a less extent in South America (see Fig. 216). Thus the greatest heat, and therefore the most effective convectional rains, as well as the convergence of the two sets of trade winds are in the northern hemisphere rather than the southern. With this displacement the typical equatorial régime, with a double maximum of rainfall, is most likely to be found between 2° S. and 8° N.

Vegetation

The vegetation associated with the equatorial climates in lowland areas is best represented by the selvas or tropical rain forests of Amazonia. This rain forest is dependent upon the maintenance of adequate and uniform precipitation and any significant departure from the normal régime, such as the appearance of a dry season, causes a deterioration of forest to grass woodland, which occurs locally within all the forest areas, but particularly in those of the Congo basin. The true rain forest is characterized by trees rising to great heights, since the prevailing temperatures and the regular rainfall promote vigorous growth. The trees develop a canopy of leaves sometimes as high as 200 feet above the ground, and the forest floor is quite gloomy (Fig. 215). This restriction of light means that plant growth on the forest floor is retarded and that undergrowth is sparse. On the edges of the rain forests, where light may penetrate more readily, undergrowth may become unbridled in its development —thus advances, made by man into these forests, have been repelled. The large trees are the hosts of parasitic plants and epiphytes (such as orchids) which lodge in the forks of branches, whilst vine-like lianas scale the giant trees and frequently link one tree to another in festoons (Fig. 231). Trees are found at all stages of growth. At any one time some are flowering, some are fruiting, whilst others may be casting their leaves. In general, the trees do not grow in pure stands, so that whilst valuable timbers are found in these forests, the practical difficulties of exploitation are great. Amongst the valuable trees are greenheart and mahogany, with rosewood and ebony of subsidiary importance.

Equatorial Areas
Congo and Amazon Basins

Both basins are in the zones of the trade winds of each hemisphere, but the Congo basin is shielded to a large extent by the East

Aerofilms Ltd.

FIG. 215.—Tropical Rain Forest, Gold Coast

African plateau, so that its rainfall is, on the whole, less than that of Amazonia, which is wide open towards the east. Most of Amazonia receives 70 inches of rain in the year, but the central part has less rain than elsewhere. The greatest precipitation is in the west (see Iquitos), and this is often attributed to the effects of the relief of the nearby Andes mountains on the trade winds. Few stations show a double maximum of rainfall, and both Para and Manaos experience two seasons in the year, a very wet period before and after the vernal equinox (known as verão), and a period of low rainfall in the late months of the year (known as inverno). South of the Amazon there is a marked maximum of rainfall in the southern summer. Beyond the Andes on the Colombian coast north-east trade winds bring ample rainfall at all seasons, since the trend of the mountains gives access to these winds. Whilst many places in the Congo basin show a double maximum of rainfall, the annual totals are less than in South America, being of the order of 50–60 inches (see Yaundé). Again whilst forest largely covers Amazonia, in parts of the Congo basin the rain forest is replaced by dry forest and grass woodlands where the annual rainfall is less and the rainfall in some months is slight.

The Guinea coastlands of Africa and the East Indies with Malaya are equatorial areas, but they have distinct monsoonal controls, and are therefore discussed on pp. 289 and 288 respectively.

(c) Tropical Climates
Main Controls

The areas which experience these climates lie immediately north and south of the equatorial areas, extending to the tropics in some instances (Fig. 216). Apart from the monsoonal climates, the facts of importance in the continental and marine types are as follows:

(a) The zones are under the continual influence of the trade winds.

(b) The regions are affected annually in the summer season on their equatorward margins by the Doldrums belt of convectional rains.

(c) Insolation is more effective than at the Equator at the time of overhead sun, since days are both longer and less cloudy.

The interaction of the first two controls distinguishes the marine type, which receives most rain from the trade winds at all seasons, from the continental type which receives most of its rain by the invasion of the Doldrums belt at one season of the year (Fig. 217). Both types are affected by convectional rainfall with the migration of the overhead sun and the related movement of the Doldrums belt into the hemisphere experiencing summer. The marine areas depend also upon the instability rainfall from oceanic air masses as they move onshore to the warm land. This rainfall may be regarded as of trade-wind origin.

Temperature

The temperature depends primarily upon the insolation, which shows seasonal variation owing to the difference between the highest and lowest noonday angles of the sun, which increases uniformly from the Equator (where the difference is $23\frac{1}{2}°$) to either tropic (where the difference is $47°$). Nevertheless, the period when the sun is nearly overhead is not so cloudy as at the Equator, and at this time the length of the day is greater than at the Equator so that the mean monthly temperatures in the hot season may be over 90° F. Cool-season temperatures rarely show monthly means of less than 70° F. It is more suitable to refer to hot and cool[1] seasons rather than to summer and winter in these latitudes. The mean annual range of temperature is thus greater than in equatorial regions, often as large as 15° F., but marine stations show much smaller ranges, as at Colon, 1·5° F., and this is largely true of places at higher levels, as at Caracas, with a range of 4° F. Since cloudiness is often very slight, daily

[1] The term 'cool', as applied to one season, is relative to the greater heat of the 'hot' season; it is always hot as judged by British standards.

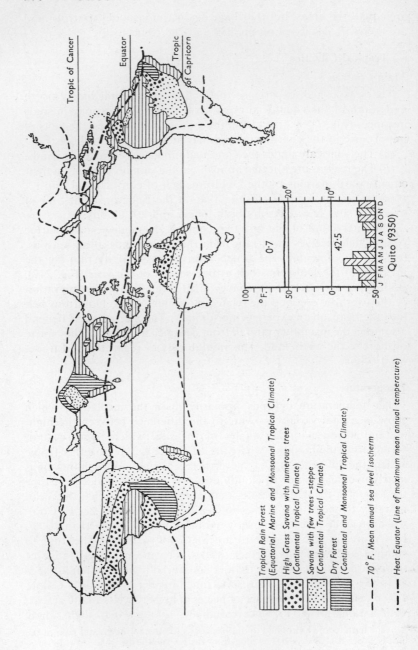

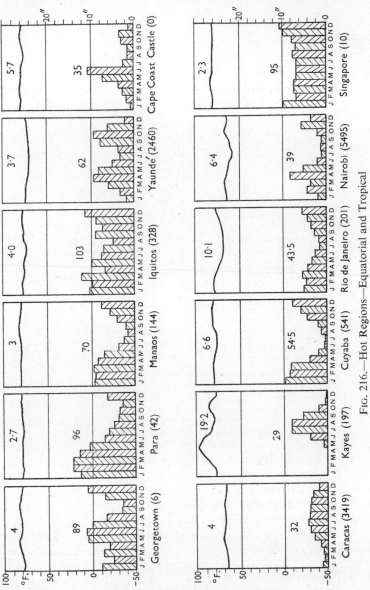

Fig. 216.—Hot Regions—Equatorial and Tropical

maxima of temperature may soar to 110° F., but these are balanced
by low temperatures at night through intense radiation, and frost
may occur.

Rainfall

The controls of the trade-wind influence and the Doldrums belt
have been mentioned. Rainfall totals are dependent on:

(a) The latitude, since the encroaching rainfall of the Doldrums
invasion in the hot season is restricted to low latitudes and decreases

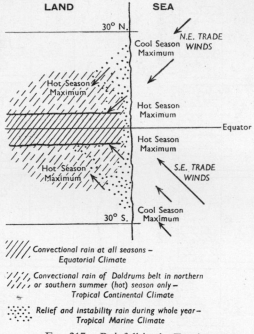

FIG. 217.—Rainfall in the Tropics

polewards, particularly in the continental interiors. This is also true
of the marine zones where the convectional rainfall may be regarded
as supplementary to the main trade-wind source.

(b) Local relief, for islands like Madagascar and Cuba have heavy
rainfall on one side and but scanty rain on the leeward or western
side.

(c) Tropical storms, tornadoes on land, and hurricanes off the
eastern coasts, may affect the rainfall totals in some areas.

There are, in the continental interiors, two seasons, one hot and rainy, the other cool and dry. The rainy season may last for seven or eight months in low latitudes (it is possibly broken by a short dry season as indicated on p. 272) and brings as much as 50 inches of rain (see Cuyaba), but this period shrinks to two months on the poleward margins, with as little as 10 inches. A double peak of temperature, apparently marking each passage of overhead sun, is shown in some graphs, but the incidence of rain in the hot season produces a marked cooling of the air and higher temperatures follow the rains (see Kayes).

In marine locations there are cool and hot seasons, both of which are rainy, but one is more rainy than the other, depending upon the relative effectiveness of the convectional and trade-wind influences. In general, marine stations show greater totals of rainfall than do corresponding places in the interiors (see Rio de Janeiro).

Vegetation

The continental interiors with their marked dry seasons are covered with tropical grasslands, although, as has been said elsewhere, 'grass woodlands' is often a more accurate description. Rain forest rarely ends abruptly, and there is a transition polewards through the grass woodlands to true savana (treeless plains). Schantz, the American geographer, has shown that, in the horseshoe-shaped zone enclosing the rain forests of equatorial Africa, the following types of vegetation may be recognized:

(a) High grass-low tree savana, which usually forms a fringe to the rain forests and is dotted with trees. The grass may rise as high as twelve feet in the wet season.

(b) Acacia-tall grass savana, having grass three to five feet in height and scattered clumps of trees.

(c) Dry forest, particularly in East Africa, which is park-like, with a grass floor, and has semi-deciduous or facultatively deciduous trees (Figs. 52 and 203).

In addition, there are areas of temperate rain forest on high ground and acacia-desert grass savana with sparse grassland and thorny trees on the poleward margins of the areas. Everywhere the trees are adapted to the dry season in having a water store, as in the case of the baobab, or by shedding their leaves (Fig. 223).

The grasses of savana areas grade from over twelve feet (Fig. 218) to under two feet with increasing latitude and decreasing rainfall. There is an immediate response to the onset of the rains, and the very rapid growth which ensues may, in the better-watered areas, result in the trees being submerged in a sea of grass. At full growth the grass is yellow and even red, but rarely is it green in colour. In the drought

Empire Cotton Growing Association

FIG. 218.—Elephant Grass, Uganda

This grass ranges up to a height of from ten to twelve feet.

season it withers and turns brown, to remain dormant until the next rainy season. Marine zones with their well-distributed and heavy rainfall have tropical rain forest, but on the leeward slopes of coastal mountains these are replaced by dry forest and grass woodland.

Tropical Regions

Africa

These regions of Africa have climates with varied régimes, but there are essentially three areas:

1. *Sudan.*—In this region the climate is almost a perfect example of the tropical continental type, having a wet season when the Doldrums belt is displaced northwards from the Equator (see Kayes). The coastal margin in West Africa is of interest (see p. 289).

2. *East Africa.*—Here the climate is essentially equatorial in régime. The graph for Nairobi shows two maxima of temperature and rainfall, but the elevation of the plateau reduces the total rainfall, so that grassland and dry forest are representative of the vegetation. The

mean monthly temperatures are relatively low, but the mean annual range is small, as with equatorial climates. Local relief produces much variety over the plateau; the rift valleys show remarkable anomalies of rainfall, for in some places forests are found, whilst in others the rainfall is so low that solid salt is preserved in dried-out lakes. The water surface of Lake Victoria increases the humidity of the local southerly winds, so that Entebbe, on the northern shore, has sixty inches of rain, which is well distributed throughout the year.

3. *South of the Congo Basin.*—The south-east trade winds give the coast a typical tropical marine climate with forests, whilst the interior only receives scanty rain from this source, but it is invaded by the Doldrums belt in the hot season. The north–south extent of these southern savanas is greater than in the Sudan, where a sharp limit is imposed by the dry north-east trade winds over the Sahara.

South America

In the north is the area of the Guianas and the Orinoco basin with a fairly normal continental régime, except on the coast, which has rain forests. On the Brazilian Plateau and its adjacent east coast there is a fairly consistent distribution of the two types of climate, as in southern Africa. Interior stations show rather greater rainfall and longer wet seasons than are usual in Africa (see Cuyaba). In North-east Brazil there is an area of deficient rainfall that gives rise to thorn forest, with acacias, known as the caatinga. Various explanations have been given for this dry region, but none is entirely convincing. It would seem that the region is sheltered by high land to the east from rain-bearing trade winds. Much of the Brazilian plateau is covered with campos cerrados (with trees) and campos limpos (without trees). It has been suggested that the term 'campos cerrados' be adopted widely for all tropical grass woodlands.

Tropical Highlands

Altitude has a marked effect on the tropical climates, as in the Andes and the East African region. In Latin America it is customary to refer to three vertical zones. Rain forests extending round the base of the tropical Andes in Peru, Colombia, and Ecuador, belong to the 'tierra caliente', which passes upwards above a level of about 3,000 feet into the 'tierra templada'. Here mean monthly temperatures are depressed to 65°–70° F. Ascending humid air gives mist and cloud, which support the quite luxuriant forest (sometimes known as cloud forest) of the Montaña in eastern Peru and Bolivia. Above about 6,000 feet the 'tierra templada' passes to the 'tierra fria', which rises to some 10,000 feet. Higher still are uplands that remain snow-covered for most of the year. The 'tierra fria' is cooler than the lower zones, and has grasslands with some dry forest. The mean monthly temperatures are remarkably uniform, as is shown by the Quito graph. In Africa the mountains of Kenya and Kilimanjaro show a similar zoning, with grass and dry forest at the base, passing up to temperate forest and, at still higher levels, to temperate grassland.

(d) Monsoonal Tropical Climates

Main Controls

There are large areas of intertropical land and seas which are under the controls known as monsoonal. The main area is in India, Southeast Asia including the East Indies, and North Australia, but there are smaller areas in Africa, Central America, and Mexico. The characteristics of a monsoonal climate may be summed up as follows:

(a) There are essentially two seasons which are characterized by opposite wind directions.

(b) At one season the winds are onshore and bring rain, whilst at the other season winds are offshore and dry or drier weather is experienced.

(c) The main criterion is this reversal of the seasonal wind directions, and in effect this means that the normal trade winds of these areas are obliterated during part of the year.

(d) The underlying cause of this reversal of the winds is the effect of continentality on temperatures and pressures in the interior of the land masses concerned.

In passing it may be noted that the word 'monsoon' means season.

India

India is the classic area of monsoonal climate, and maps in Fig. 219 show the general seasonal conditions in the great theatre in the East, of which India is only a part. India is in some respects self-contained in that the plateau of the Deccan and the northern lowlands lie to the south of high mountains which form a climatic barrier. The seat of the controls is the plateau region of Inner Asia which, in winter, is covered by cold air, owing to the intense cooling, and this is reflected in the high pressure values. However, cold air does not descend into India, whose January temperatures are never low, ranging from 80° F. in Ceylon to 50° F. in the north. Nevertheless, there are outblowing winds over the country as a whole in this season. In summer, one of the most extensive hot zones of the world stretches from Africa over India and on to the inner plateaus of Tibet and Mongolia. This zone

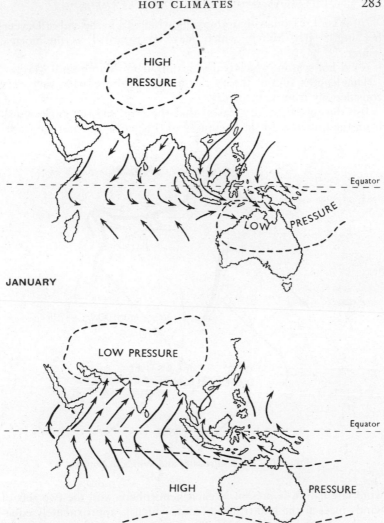

Fig. 219.—Pressure and Winds in Southern Asia and Australia

replaces the winter high pressure by markedly low-pressure air, which has its deepest centre in the Sind region of North-west India.

The Indian year may be divided climatically into three parts:

(*a*) A hot season (with inblowing winds from the sea that bring rain) which extends from June to mid-September in the north, and from June until mid-December in the south.

(*b*) A cool season, with outblowing winds, from the end of December until early March, which may be regarded as the normal winter.

(*c*) A hot season, which is dry, lasting from March until May.

The lengths of the three seasons given above may vary very considerably from year to year.

It is during the hot dry season that the overhead sun moves north from the Equator. At the beginning of this period the normal trade-

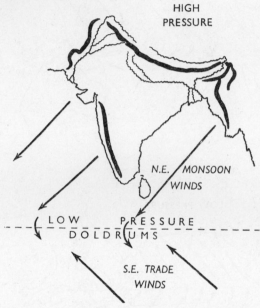

FIG. 220.—India—Pressure and Winds in March

wind circulation is present in each hemisphere, and the two sets of winds meet at the intertropical front, which is approximately equatorial in its position (Fig. 220). The culmination of this season brings a very different set of conditions (Fig. 221). The low pressure in India has reached a critical depth, and the Doldrums belt of low pressure near the Equator may be considered as having relatively high pressure. There is, in fact, a pressure gradient from the southern hemisphere, and the south-east trade winds transgress the Equator into India.

The rainfall that follows is ushered in by the 'burst' of the monsoon which is preceded by days and even weeks of very variable winds and intermittent storms of the hurricane type, known as cyclones in the Indian Ocean. The onset of the south-west winds of the wet or summer monsoon is usually accompanied by massive clouds, and the impact of warm tropical maritime air with hot land surface results in torrential instability rainfall, which is increased locally by high relief. On the flanks of the Western Ghats annual rainfall of over 200 inches is common, but a rapid decline to 20 inches is usual to the east of the crest of the Ghats. The extensive rain zone that lies over the eastern part of the Ganges plain and in Assam shows diminishing totals towards the Sind focus (Fig. 221). Probably the enclosing mountains help to produce the deflection of the winds from south-west to south-east in the north of India, but the indraught to the low pressure of the Sind is of paramount importance. The deficiency of rain in the Sind is surprising but, high above the area, hot dry winds from Baluchistan effectively restrict condensation in rising air and rain is unlikely. One feature of the rainy season in most parts of the country is the extremely high humidity which, accompanied by high temperatures, is both irksome and physiologically dangerous to Europeans, who resort to hill stations whenever possible.

The end of September marks the coolness that is consequent upon the return of the overhead sun to the southern hemisphere. The low pressure over India apparently retreats southwards with the establishment of high pressure in the interior of Asia. This change resembles the retreat of the convectional rains in the continental interiors of Africa, and it is therefore feasible to consider the low pressure that covers northern India in summer as the Doldrums belt which has been displaced northwards to an extraordinary degree. The ensuing season is cool, and apart from some regions this is a time of drought. Rain falls in eastern India, particularly on the south-east coast of the peninsula and in Ceylon, since north-east winds of the winter monsoon pass over a water surface in the Bay of Bengal (see Madras). North-west India also receives rain at this time from depressions that come from the west, probably from the Mediterranean Sea basin (see Lahore). After the winter solstice, with the northward march of the overhead sun, the cool season merges with the hot, dry season that leads to the following year's monsoonal 'burst'.

The vegetation of India is mainly tropical rain forest in the wettest parts and monsoonal forest elsewhere, apart from the drier north-west, where desert and steppe grassland prevails. Monsoon forest consists of semi-deciduous or facultatively deciduous trees, amongst which are sal and teak, both exhibiting xerophilous characteristics in the dry season. However, the Deccan has large areas of desert grass savana.

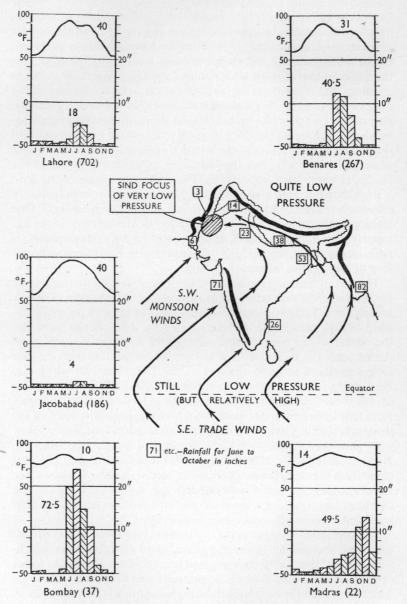

FIG. 221.—India—Pressure, Winds, and Rainfall in the Wet Season

North Australia

This area is shown by Fig. 219 to be monsoonal, with the reversal of winds in relation to seasonal changes in the continental pressure. Unlike India, it has no 'burst' of rains heralding the wet season, when the north-east trade winds transgress the Equator and become north-west winds blowing into a focus of low pressure in North Australia. The rains are heaviest on the coast, and extend inland as shown by Fig. 222. The dry season of winter has outblowing winds from the

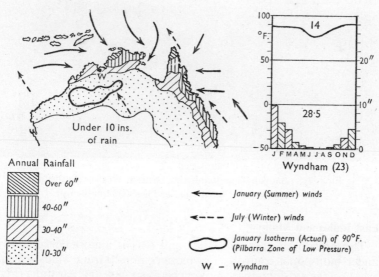

FIG. 222.—North Australia—Winds and Annual Rainfall

south-east. On the Queensland coast there is rainfall from onshore trade winds at all seasons. Coastal hurricanes and storms, known as willy-willies, bring additional rainfall to the north. These parts of Australia may be regarded as having a tropical continental climate, with a marine climate in the east. On the north coast a narrow forest belt passes inland to savana with scrub (known as savana woodland in Australia) (Fig. 223), and eventually merges with the interior desert. In Queensland the east coast is forested, but local relief may produce a deterioration of rainfall, so that grass woodlands appear. The predominant forest trees are eucalypts. There is also a eucalypt scrub called mallee. Acacias are also common in the interior grasslands, and produce a scrub called mulga and, in some parts, brigalow.

Black Star

FIG. 223.—Baobab Tree, Kimberley District, Western Australia
Note the grass in the foreground and woodland in the rear.

East Indies and Malaya

This equatorial area of islands and seas is almost continuously under monsoonal controls, and receives rain-bearing winds whichever way the air streams are moving between Asia and Australia (Fig. 219). The climate is probably most suitably described as monsoonal equatorial, because its régime shows a mixture of the two types, in that stations usually show a double rainfall and temperature maximum, but the rainfall maxima are primarily due to wind influences rather than to increased convection with overhead sun. On the other hand the climate might be regarded as tropical marine, since the well-distributed rainfall depends on trade winds. The region receives an influx of warm, moist air at all times of the year, although when the winds are blowing from Australia the rain is rather less. Rainfall totals are rarely less than 80 inches per annum (see Singapore), and often more where high relief intervenes. The two rainy seasons are separated by less rainy periods, which mark the change-over of the pressure gradients from Australia to Asia and vice versa. Temperatures are equatorial in magnitude, and the mean annual range is very small indeed (see Singapore). Much of the area has tropical rain forest, whilst on high land savana is found.

Other Areas

In Africa there are two other regions with monsoonal climates. As Fig. 219 will show, Abyssinia is, in effect, an extension of the Indian area. Summer monsoonal rainfall here is responsible for the annual floods of the Blue Nile. The coastlands of the Gulf of Guinea may also be regarded as monsoonal. In general, the Doldrums belt and the inter-tropical front remain for most of the year along these coastlands. Dry north-east winds (harmattan) blow from the interior, and south-west winds, which are extensions of the south-east trade winds into the northern hemisphere, blow onshore and are responsible for the abundant rainfall. However, the north-east winds may bring a dry season to parts of the coast. At one point the configuration of the coast produces an upwelling of cold water (cf. Peruvian coast, p. 293) which reduces the rainfall on land, so that locally grass woodland replaces the rain forest. Some areas have a double maximum of rain (equatorial régime), whilst others have a marked wet season in the northern summer, when the inblowing south-west winds are most effective (see Cape Coast Castle). Mangrove swamps fringe this coast, and these may be regarded as extensions of the land forests on to the sea. Finally, a brief mention of Central America and the Caribbean area may be made. Whilst north-east trade winds prevail throughout the year, there is in the hot season an indraught of rain-bearing winds to the islands and the mainland because of low pressure. The warm waters here induce this low pressure and in summer winds may blow from the south-west on the west coast of Central America. In all seasons local aspect plays an important part in determining the rainfall. Windward coasts are clothed with tropical rain forest and at high levels there are regions of cloud forest.

(e) Hot Deserts

The normal positions of the hot deserts are on the poleward margins of the tropical grasslands, within the trade-wind belt and athwart one of the tropics. They usually extend to western shores, but rarely to the eastern shores of land areas. The aridity of the deserts depends on the trade winds, which are dry, having blown far inland from the oceans. Furthermore, these areas usually have the high pressure of the subtropical cells which are associated with low rainfall. Even on their western margins the proximity of the ocean is of no avail in relieving desert conditions, since any inblowing winds can give very little rainfall because of the high temperatures on the land combined with the effect of cold currents offshore.

Temperature

The maxima of temperature in the hot deserts are certainly as high as anywhere in the world. A minimum of cloud cover gives unrestricted insolation, which is increased in its effect by the almost complete absence of water surfaces and of vegetation. In the hot season daily maxima of 130° F. are common, but the great loss of heat at night, owing to the very great radiation with cloudless skies, may produce freezing temperatures. Diurnal ranges may be as great as 60° F. or 70° F. The mean annual range of temperature is usually less than 30° F., with the mean monthly temperatures varying from about 80° F. to 50° F., although higher maxima are common in the interiors (as at Insalah, whose hot-month temperature is 100° F.). Compared with the monotony of the temperature in equatorial climates, it will be appreciated that the deserts enjoy relatively invigorating climates.

Humidity

The aridity which is characteristic of deserts may be defined in terms of the relative humidity of the air. Values of 50 per cent. and below are quite common and, in some extreme instances, 10 per cent. or less has been recorded. It may be of interest to note that measurements of the absolute humidity may show that there is, in fact, a fairly large water-vapour content in the desert air, but with the prevailing high temperatures of the daytime, much water would have to enter the air as vapour before it could become saturated. In view of the general absence of water surfaces from which evaporation may proceed, it is virtually impossible for high values of relative humidity to be recorded in the daytime except on western coasts. Dew and hoar frost may occur at night.

Rainfall

It is not so much the low annual rainfall as its incidence which determines a desert. For instance, desert may prevail where about fifteen inches of rain falls in the year if this precipitation is restricted to a few torrential downpours. On the other hand, it is possible for crops to flourish in areas with less than ten inches of annual rainfall provided that sufficient falls in the vegetative season of the plant concerned. In general, it may be assumed that desert climates are experienced where the mean annual rainfall is less than ten inches. Within the interiors of the hot deserts there is so great an uncertainty of rainfall that it is generally true to say that rain is negligible: in several years only one storm of rain may occur. Torrential showers may not be unusual in some areas, but their duration is limited and their effects are short-lived. However, in some deserts it is possible to distinguish definite seasons of rainfall.

The vegetation of desert regions is discussed on p. 249. There is rarely, if ever, a clearly marked boundary to a desert. In most cases a transition through semi-desert, with scanty seasonal rainfall, may be traced from north and south. On the equatorward side this transition zone has hot-season rainfall (related to the savana areas), whilst on the poleward margins the rain has a winter incidence (related to Mediterranean areas). In passing, it may be noted that there are very few areas of waterless desert, as will be seen by a reference to Fig. 224, which shows the world distribution of hot deserts.

Continental Areas

In Africa there is a great difference between the east–west breadth of the deserts of the north and south. In the south the south-east trade winds blow from sea to land, whereas in the north the prevailing north-east trade winds arrive at the Red Sea coast of Africa as dry winds, having blown from interior Asia with little or no chance of gaining water vapour as they cross the sea from Arabia. The south of Africa has a well-watered east coast, so that in conjunction with the narrowness of the continent in these latitudes, desert is relatively restricted. In the north, aridity prevails over the very wide continent and, indeed, the Sahara must be regarded as part of a greater desert zone which continues eastwards into Arabia and Iran. In Australia the controls which produce the hot desert are similar to those in southern Africa, but the greater east–west extent on the continent gives a correspondingly wider area. North-west monsoonal winds blow in the hot season from the seas to the north of Australia, but they have only marginal influence as regards rainfall, although they are attracted inland by the seasonal low pressure. Here conditions recall those of the Sind region in India (p. 285), where in the hot season little or no rain falls. In North America the hot desert is very limited on the narrow Mexican plateau.

Marine Areas

In Peru and northern Chile the hot deserts are entirely marginal, and a study of local controls will be fairly typical of those prevailing in similar areas on the west of the other hot deserts. In Fig. 225 the essential controls are shown. The trade winds are offshore, and are related to the high-pressure cell in the south Pacific Ocean. These winds are partly responsible for the current of cold water which flows

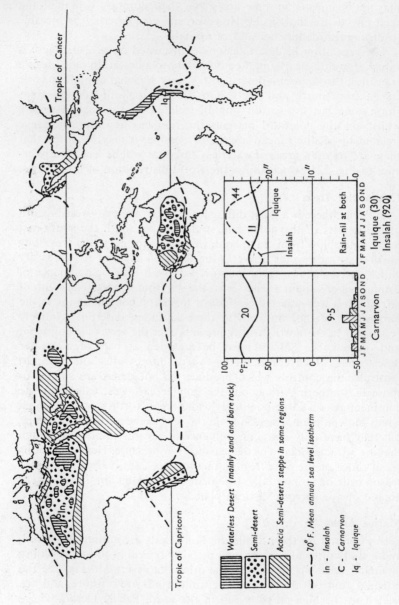

Fig. 224.—Hot Regions—Hot Deserts

northwards along the coast, producing an upwelling of deep ocean water which is much cooler than the surface water (p. 347). The desert areas of western South America show continentality in the development of low pressure in the hot season, which causes a movement of air towards the shore from the ocean. The moist oceanic air is chilled in passing over the water of the cold current, and condensation occurs in the form of mist, which extends on to the coastal lands, where it is dispersed by the warmth of the land, so that not far inland the typical desert climate is preserved. The mist (known locally in Peru as the garua) often produces a gloom, accompanied by the dripping of moisture from buildings and trees.

The moist cool air reduces the hot-month temperatures, and the mean annual range of temperature is smaller by some 10° F. than that

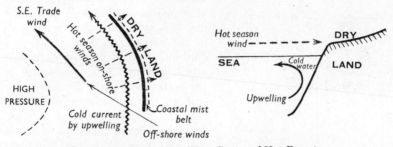

FIG. 225.— Controls on West Coasts of Hot Deserts

experienced in interior desert areas (see Iquique). Locally, hot dry air may flow from the interior, and give much higher temperatures at the coast. However, in general, the mean monthly temperatures range between 55° F. and 70° F. and show the usual equability of marine zones. Rainfall is very scanty, and generally coincides with onshore winds in the hot season.

The Peruvian-Atacama Desert is a marine hot desert, and is very narrow from east to west, since the high wall of the Andes mountains almost completely isolates the area from continental and eastern influences. The cold current helps to strengthen the aridity, and this is the hot desert with the greatest north–south extent in the world. It stretches from just south of the Equator to lat. 30° S. The importance of the cold current is shown in the northern part of Peru, where copious rainfall may interrupt the desert régime approximately every seven years (sometimes it is after a fourteen-year interval), when a warm current of water[1] sweeps southwards from the northern

[1] This current arrives off Peru about Christmastide, and is known as El Niño (the boy child).

hemisphere and completely counteracts the influence of the cold water, so that moist onshore winds produce the very heavy rainfall on the land.

Similar controls are found in South-west Africa and elsewhere, but in Western Australia there is a very short desert coast, partly because the cold current offshore is much less effective than the currents along other west-coast deserts in the hot regions (see Carnarvon).

(f) Soils of Regions with Hot Climates

Associated with the climatic regions which have been reviewed, apart from those in desert areas, most of the zonal soils are predominantly red in colour and usually show a high degree of laterization. Both the tropical rain forests and the continental savana areas are widely covered with such soils, although savana regions with their dry seasons have other zonal soils of the black-earth type, in which a concentration of calcium carbonate affects the colour of the humus. In the Sudan and in the Deccan, some soils are rich in useful minerals when they have a parent of lava rock. Intrazonal soils occur as well, and amongst these are the grey and black clays of the forest swamps.

With the various red soils are also yellow and sometimes almost white soils, the products of laterization.[1] A typical profile is shown in Fig. 226. The soils are not very fertile, as the humus content is low, and there is marked acidity, since many alkaline salts are eluviated from the upper horizons. They are usually of good structure and well drained. In the light-coloured soils most of the iron oxides are leached out, leaving much alumina at the surface, which is whitish in colour, but often marked red with some remaining iron oxides.

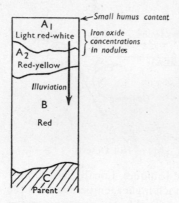

FIG. 226.—Profile of Tropical Red Soil (Lateritic)

The tropical grassland soils have dark surface horizons, with much humus, and are somewhat more valuable. They are, however,

---

[1] Laterization removes practically all but iron and aluminium oxides, so that the upper horizons are left with a deficiency of silica. Most tropical red soils have an $\frac{SiO_2}{Al_2O_3}$ ratio of more than 2 near the surface, but in some cases this ratio is less than 1·33, when the clay so formed is truly laterite or sometimes bauxite, the latter being almost pure alumina.

rapidly exhausted by cropping, and soil erosion is readily set in motion when the grass cover is disturbed. Some savana areas have soils which compare with the chernozems (see p. 330), but it is questionable whether they are truly of this type. Under cooler conditions in the tropical highlands podzols are found.

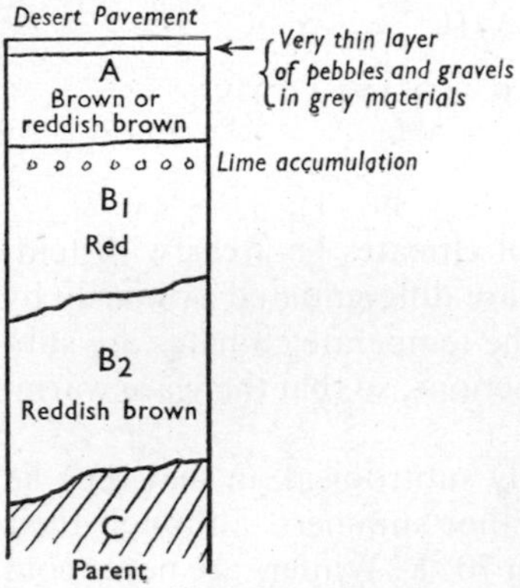

In the tropical deserts the main zonal soils are grey. They have little organic material and light colours are common at the surface, but digging reveals the brighter red and yellow layers below. Drought produces an upward movement of various salts, and there is usually a zone of lime accumulation at, or just below, the surface, and in some instances the surface is quite saline when only halophytes (salt-loving plants) will prosper. With irrigation, as is possible on alluvial fans, excellent crops may be raised, as in the Peruvian desert. The profile of a desert soil is given in Fig. 227. It must be realized that much of the desert regions is unlikely to possess true soils in view of the bare rock surfaces and the sand dunes.

FIG. 227.—Profile of Desert Soil (Arizona)

CHAPTER XVIII

WARM TEMPERATE CLIMATES

Introduction to Temperate Climates

Polewards from the regions with hot climates lie areas with temperate climates. Whereas hot climates are differentiated primarily by rainfall and its seasonal distribution, the temperate climates are subdivided primarily by temperature distinctions, so that there are warm, cool, and cold types.

Warm temperate areas are essentially subtropical, in that they lie just outside the tropics and have very hot summers, although their mean annual temperatures are less than 70° F. Winters are never cold enough to arrest all plant growth, and never does the mean temperature of the coldest month fall below 43° F.

Cool temperate climates have at least one month (on the equatorward margins) and often five months (on the poleward margins) with mean temperatures below 43° F. Thus, a rest is imposed on the growth of plants. Cold temperate climates have longer cold seasons, with at least six months with mean temperatures below 43° F.

The cool and cold temperate areas lie wholly within the westerly wind belt with its associated depressions, whereas warm temperate zones are alternately under trade-wind and westerly wind influences during the year, since they lie at a significant meteorological boundary, namely, the Horse Latitudes.

The warm and cool temperate climates are each capable of a subdivision into west marginal, interior, and east marginal types. Both the margins have rainy seasons during the year, but the interiors are quite dry, and include cool or temperate deserts. This is particularly true of the northern continents, where the continentality of the interior produces this triple division. Cold temperate regions are not so clearly divided, as desert does not appear in the interior areas. In Asia the eastern margins of all the temperate areas have monsoonal controls, which also affect the east marginal areas in the warm temperate latitudes of North America.

Distribution of Warm Temperate Climates

Lying mainly between 30° and 40° N. and S., these regions are under the influence of the trade winds and westerly winds, with depressions, in alternate seasons of the year. Fig. 228 shows the essential contrast between the west and east margins. The former have offshore winds in summer and onshore winds in winter, the rainy season. The latter have onshore winds in summer and outblowing winds in winter,

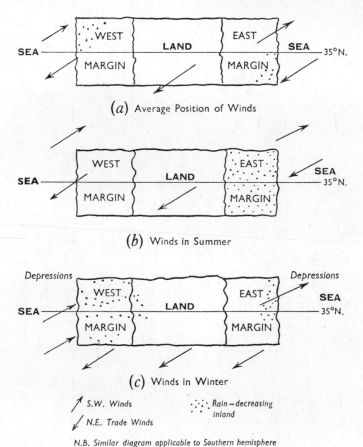

(a) Average Position of Winds

(b) Winds in Summer

(c) Winds in Winter

↗ S.W. Winds   ∴ Rain – decreasing
↙ N.E. Trade Winds   inland

N.B. Similar diagram applicable to Southern hemisphere

FIG. 228.—Winds and Rainfall in Warm Temperature Areas
(Northern Hemisphere)

but both seasons are rainy, as will be seen later, although summer is the more rainy. West and east marginal areas are found paired with one another in all continents (see Fig. 229). In South Africa the two marginal areas are practically contiguous, but elsewhere there is a gap of varying width between the margins which is occupied by fairly arid regions.

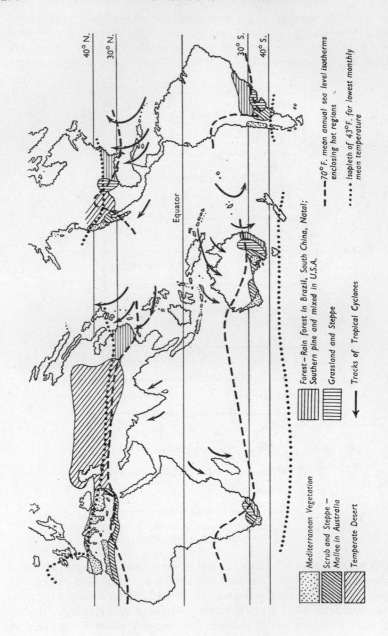

Forest – Rain forest in Brazil, South China, Natal;
Southern pine and mixed in U.S.A.

Grassland and Steppe

Tracks of Tropical Cyclones

– – – 70° F. mean annual sea level isotherms
enclosing hot regions

· · · · · Isopleth of 43°F. for lowest monthly
mean temperature

Mediterranean Vegetation

Scrub and Steppe –
Mallee in Australia

Temperate Desert

40° N.

30° N.

Equator

30° S.

40° S.

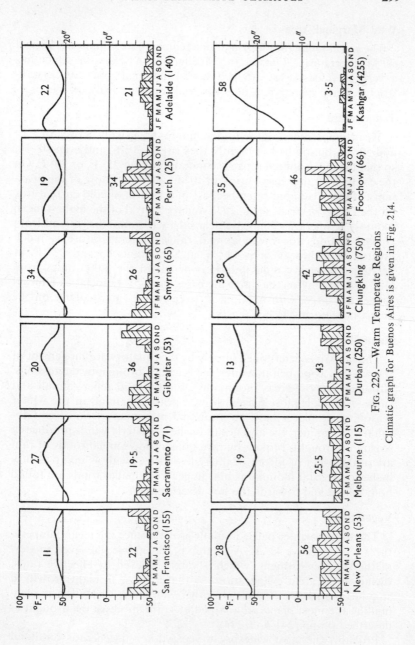

FIG. 229.—Warm Temperate Regions

Climatic graph for Buenos Aires is given in Fig. 214.

West Marginal Areas

The warm temperate west marginal climates are commonly known as 'Mediterranean'. Outside the Mediterranean Sea basin, which has a very great east–west extent, the west marginal zones are very restricted, being merely coastal strips and therefore truly marginal.

Temperature

Mean temperatures of the coldest months are usually between 43° F. and 50° F., but in the hot months they rise to 70° F. and even 80° F., so that the mean annual range is commonly about 25° F. or 30° F. On the coasts the range may shrink to 10° F., partly because of the proximity of the ocean, but often because of the effect of cold currents (as off the Peruvian coast, see p. 293), and the arrival of the highest mean monthly temperature may be delayed by two months. Inland from coastal stations the mean annual range of temperature is much greater. In California it is 11° F. at San Francisco and 27° F. at Sacramento, eighty miles inland, whereas in the 2,000 miles from Gibraltar to Smyrna the range increases from 20° F. to 34° F. Diurnal ranges are large, with maxima of over 100° F., particularly on the desert margins in the hot season.

Rainfall

The rainfall is determined largely by the equatorward movement of the westerly wind belt in winter and is mainly brought by depressions, although in most areas mountains produce local relief rainfall and snow. Although most places show a peak of rainfall in the winter season, some places have an autumnal maximum of rain (see Gibraltar). In summer, the absence of onshore westerly winds produces a season of drought, broken by rare rainstorms. Annual totals of rain are of the order of fifteen to thirty-five inches, with as many as fifty inches on coastal mountains but, in general, smaller totals are found both inland and towards the hot desert margins.

Vegetation

There is no winter pause in plant growth, since the mean temperatures remain above 43° F. A rhythm is perceptible, however, in view of the summer drought, which is rarely relieved by effective rains, owing to the great evaporation, so that a pause in plant growth in summer is almost inevitable. In other words, the vegetation is xerophilous and most plants are adapted by nature to resist the drought as described on pp. 241 and 244.

Probably the most widespread vegetation is degenerate woodland in the form of evergreen shrubs and thickets which are practically impenetrable. In the several 'Mediterranean' regions such plant associations are so common that local names have arisen, such as maquis (France), macchia (Corsica), mallee (Australia), and chaparral (California) (Fig. 230).

In areas with scanty rainfall, and on pervious limestone in better-watered regions, the characteristic vegetation is very sparse and is known as garigue in the Mediterranean Sea basin. Garigue consists of small, brightly coloured, sweet-smelling shrubs and plants, such as

Dorien Leigh Ltd.

FIG. 230.—Macchia on the Coast of Corsica

thyme, lavender, rosemary, rock-rose, and broom, with bare rock and soil lying between the separate plants. Grass is usually very restricted, because its short roots are unable to draw upon deep moisture in the warm dry season (Fig. 123).

Mediterranean Areas

1. Mediterranean Sea Basin

This great embayment is quite exceptional amongst 'Mediterranean' regions with its east–west structural trends which guide depressions passing in winter between the high-pressure areas of the Azores and the European continent. Annual rainfall is greater on west-facing coasts than on the eastern sides of the peninsulas of the European shore, but decreases on the whole gradually from about thirty-five inches at Gibraltar to slightly less than twenty-five inches on the coast of Palestine, over 2,000 miles farther east. Cold winds, such as the bora and the mistral (p. 222), blow in the rear of depressions, and warm winds drawn from the Sahara, such as the sirocco (Algeria) and the khamsin (Egypt), arrive on the European coasts both humid and warm.

In summer the basin is covered by an extension from the west of the Azores high-pressure zone, which effectively shuts out depressions, although convectional rain-storms may occur. Etesian winds, from the north, blow in the eastern part of the basin on to the Libyan coast, giving fogs. The peninsulas of Iberia, Italy, and Greece exhibit continental features of climate which make them exceptional, as is reflected in the steppe vegetation of the Meseta.

2. California and Chile

These two areas of the New World are alike in having north–south trend lines, with an interior central valley backed by high mountains and separated from the sea by coastal ranges, occasionally broken by gaps, such as the Golden Gate in California. Cold currents offshore give cool air, which is drawn in summer through the coastal gaps into the valley that is a local centre of great heat and low pressure. Fog is a feature of the coasts in summer, and until the indraught of cool air ceases in late summer, the highest mean monthly temperatures of 60° F. and 67° F. are not recorded at San Francisco and Valparaiso, respectively. Winter rain, which is somewhat variable in amount, is brought by depressions, which often pass across the coast on the poleward side of these areas. Warm tropical air drawn north in California gives the dry wind known as the Santa Ana. Aspect and altitude have a bearing on the rainfall totals, for the valley floors are less well supplied than the seaward-facing slopes of the ranges, although the maximum precipitation is below the crest of the interior mountains. Redwood forests flourish in the northern coast ranges and on parts of the Sierra Nevada in California, but the usual vegetation is degenerate woodland or chaparral in California (Fig. 209).

3. Cape Region and Swanland (with the Adelaide District)

Both these areas lie on the southern margin of a continent and are backed by plateaus. Throughout the winter the proximity of the sea to the west and the south prevents very low temperatures. Perth, in Western Australia, is practically frost-free, and its lowest mean monthly temperature is 55° F. The Berg wind, which is föhn-like, descends from the plateau in the Cape region in winter, giving day temperatures as high as 100° F., but there is no counterpart of this wind in Swanland. Depressions crossing the seas to the south bring very reliable winter rainfall, as well as some in summer, but this declines inland. Eastwards from Swanland the northward curve of the Great Australian Bight causes rain-bearing winds to miss the land. Farther east the coast trends southward, and in the Adelaide district and parts of Victoria the rainfall increases: there is no summer drought here (see Adelaide), but there is clearly a winter maximum of rain. In summer temperatures are moderated (slightly) by the proximity of the ocean and also by a cold current off the Cape region in particular. The areas have a succession of vegetation related to the decrease of rainfall inland. With forty inches of rain on the southwest coast, Swanland has eucalypt forests of jarrah and karri which pass to mallee inland.

East Marginal Areas

The eastern areas in warm temperate latitudes have rain at all seasons, although summer is very often the wetter half of the year with onshore trade winds. However, both temperate and tropical cyclones affect the areas. In winter temperate depressions, with their contrasted winds, may produce rapid changes in the weather which are characterized by sudden falls of temperature through 30° F. to 40° F. in a few hours.

Temperature

Despite the possibility of very cool weather, the mean annual range of temperature is generally small. The oceanic influence on the southern hemisphere regions may be contrasted with the continentality that affects the northern areas. In the south, mean monthly temperatures range from 50° F. to about 70° F. (occasionally rising to 75° F.), whereas in the north a mean annual range of 30° F. or more is common with a hot-month temperature of over 80° F. (see New Orleans). In summer the high humidity of the onshore winds gives wet-bulb temperatures of over 70° F., and thus the activities of white people are restricted in these areas.

Rainfall

In summer generous and heavy instability rainfall is produced by moist, warm, oceanic air moving on to the heated land surfaces. Locally this rainfall is augmented by relief, and also by typhoons and hurricanes. Winter rainfall is often brought by depressions. Almost everywhere near the coasts the annual rainfall exceeds forty inches, and often fifty inches, although the totals decrease inland. Summer is not everywhere the wettest season, and in any case evaporation tends to reduce the effectiveness of the summer rain.

Vegetation

The mild and even warm winters permit the perennial growth of plants, and since there is no marked dry season in the year, forests are common, especially near the sea margins. Trees are similar to those of the west margins, with conifers (cypress), laurels, and evergreen oaks, but in addition there are the magnolia, the mulberry, the maple,

Prof. E. Ashby

FIG. 231.—Forest in South Carolina, United States of America

This forest resembles the tropical rain forest. The trees are oaks and they are festooned with lianas.

the walnut, and the hickory, which form broad-leaved evergreen forests. Lianas and bamboos occur in some areas, and recall the tropical rain forests (Fig. 231). Deficiency of rainfall in the interiors results in the formation of scrub, and eventually this passes to grassland and steppes.

Southern Regions

It is convenient to discuss the east margins in the southern hemisphere separately from the northern areas. This is because of the marked continentality which gives monsoonal characteristics to the latter areas, particularly in China. Furthermore, tropical cyclones are important in the northern regions, whereas they are less frequent in the southern hemisphere.

1. New South Wales and Victoria

Much of the winter and autumnal rainfall (see Melbourne) is produced by depressions which pass eastwards through Bass Strait, although they may penetrate northwards. Troughs of low pressure, particularly in summer, separating migrant anticyclones give brickfielders and southerly bursters. The former are hot winds blowing from the continent and are accompanied by much dust, whilst the latter are quite cool and violent blasts from the south, usually marked by a dark roll cloud associated with a cold front. The burster often brings, despite its coolness, relief from the suffocating heat of the brickfielder (Fig. 232). Annual rainfall decreases from twenty-five to forty inches and more on the coast, where

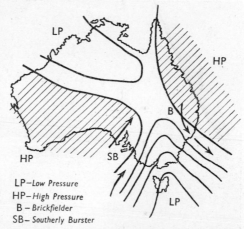

LP—Low Pressure
HP—High Pressure
B—Brickfielder
SB—Southerly Burster

FIG. 232.—Pressure Distributions in Australia with a Trough of Low Pressure giving Southerly Burster

forest occurs, to under twenty inches inland, where grassland and scrub flourish on the highlands and the interior plains.

2. South-east Africa with Natal

As progress is made along the coast from Cape Town towards Durban, there is an almost perfect transition from the west to the east marginal warm temperate climate. Knysna is often regarded as marking the boundary, with more rain in winter to the west and with a summer maximum of rainfall to the east. Durban, with the moderating effect of the sea, has very warm winters and a mean annual temperature of 70° F., which places it on the edge of the tropical marine zone of East Africa. Its mean annual range of temperature is 13° F., between 64° F. and 77° F. No cold winds of polar origin are experienced, although a föhn wind may elevate the diurnal temperatures in winter. Decreasing rainfall towards the interior results in a change from forest to steppe, with grassland on the Veld (Fig. 233) and dry forests of wattles (acacias) in some areas.

3. South America—Central Argentine and Uruguay, with Southern Brazil

Migrant depressions in winter passing eastwards bring the zonda, a warm, humid wind from the north and the cold pampero from the south (a squally wind but not so vigorous as the southerly burster of

Black Star

FIG. 233.—Veld Plateau in Orange Free State

The vegetation is warm temperate grassland. Note also the plateau surface, which is broken by mesa-like hills, and the strato-cumulus clouds.

Australia), as well as south-east winds off the sea to the east. In summer low pressure develops over the interior and splits the sub-tropical high-pressure belt into two oceanic cells to the east and west of the continent. Onshore winds from the north-east, east, and south-east blow inland from the Atlantic Ocean and account for the summer rains. The annual rainfall totals decrease steadily inland, an effect of the subdued relief of the region, but only in Southern Brazil is there forest, elsewhere are grasslands. However, Schmieder has advanced the theory that an earlier forest cover on the Pampas has been destroyed by burning and clearance. Local scrub or 'monte' supports this view, and wind-breaks on estancias prove that trees can flourish in the area. The climatic graph for Buenos Aires is given in Fig. 214.

Northern Regions

The 43° F. mean January isotherm is taken as the poleward limit of the warm temperate climates, and this lies close to the River Yangtse Kiang, so that North China, despite its warm summers, has quite cold winters (see p. 320). The same isotherm marks, by chance, the northern edge of the North American cotton belt. Both South and Central China, as well as the Gulf-Atlantic States, are monsoonal, although winter depressions tend to mask the seasonal reversal of the wind directions, particularly in North America. Tropical cyclones are common in both of these northern areas.

1. South and Central China

In winter, the normal north-east trade winds are reinforced by northerly outblowing winds from the high-pressure zone of inner Asia, from whose southern margin almost inexplicable depressions emerge into China, and bring rainfall to the Yangtse Kiang valley. Behind these disturbances polar air may surge southwards and give freezing temperatures in Canton. Nevertheless, Szechwan (see Chungking), in the track of these depressions, has extraordinarily mild winters, with January mean temperatures about 11° F. higher than has Shanghai. In summer, as temperatures mount, there is a period somewhat like the hot, dry season in India, but rain is not entirely absent, and these is no 'burst' of rain when the south-east winds set in from the sea towards the continental low pressure. Much of this summer rainfall is apparently brought by local cyclonic disturbances. Mean monthly temperatures may surpass 80° F. for several months, and with the great humidity the climate is almost equatorial.

Typhoons in late summer and autumn affect ports between Hong Kong and Foochow, the latter showing a distinct peak of rainfall in August (see graph). With their recurvature around the western margins of the North Pacific high-pressure cell, Shanghai enjoys a relative freedom from these storms.

Fifty inches of rain, and even more at coastal stations, are usually well distributed throughout the year. On the south coast rain forest occurs but is replaced northwards and inland by broad-leaved evergreen forest, except where clearance has been made.

2. Gulf-Atlantic States of North America

Here monsoonal controls are weaker than in China. Depressions from the Pacific Ocean follow tracks as shown in Fig. 239, and bring winter rainfall, with the tropical air drawn off the Gulf of Mexico. Behind these depressions the cold northers, in Texas and farther east, may lower temperatures by 50° F. in a few hours. In summer the indraught of warm humid air from the south gives high temperatures and great humidity over a large part of the Mississippi lowlands; 80° F., with 80 per cent. relative humidity, is frequently recorded as far north as St. Louis. Summer is the season of abundant rainfall and most stations show a maximum then, despite the considerable precipitation in winter from depressions. On the coasts hurricanes may give heavy autumnal rainfall, but these storms are less numerous than the typhoons in China. Although much land is under cultivation, there are still large tracts of virgin forest, including the long-leaf southern pine belt, and on the coast in Florida are the Everglades, with swamps and almost impenetrable woodland.

Temperate Deserts

In Eurasia there is a wide intermediate zone between the east and west margins in warm temperate latitudes which is occupied by desert that may be regarded as a northerly extension of the hot deserts. A similar but much more restricted area is found in North America. The more northerly position, the elevation, and the rather special physical nature of these regions make them different from the tropical deserts.

They are cooler than the hot deserts, having in Asia January mean temperatures that often fall below 32° F., partly because of the elevation above sea-level, but also because of the effects of continentality (see p. 164). However, the summers are hot, and the lack of cloud gives daily maxima of temperature equal to those of the tropical deserts, that is, frequently greater than 100° F. Much of these areas consists of upland basins shut in by high mountains which isolate them from rain and snow in winter. In any case as these are centres of very high-pressure air in winter the chance of precipitation is slight. The dryness of the regions in winter is reflected in the rapid rise of temperatures in spring; little insolation is required for the snow melt prior to the heating of the ground surface.

Rainfall is slight but mainly of summer incidence because of the continental régime (see p. 234). The climatic graph for Kashgar is typical. Some winter rainfall is received in the Eurasian zone from occasional depressions which have come from the Mediterranean Sea basin, although such rainfall is mainly in the western parts.

The vegetation is very commonly steppe (Fig. 234), or sage brush (in North America). On the ranges surrounding the intermontane basins precipitation may be relatively great, so that grasslands and even woodlands are found. Polewards, in North America, and farther into the interior of Central Asia, these deserts persist, although in some parts mountain climates with their special features are found.

Literary Services Mondiale Ltd.

FIG. 234.—Steppe in Mongolia

The vegetation consists of very sparse grass in an intermontane plateau region of Central Asia.

Temperate deserts extend into cool temperate regions and are discussed in the following chapter.

Soils of the Warm Temperate Regions

The threefold climatic division is reflected in the soils. The east margins, with their well-distributed rainfall, have red and yellow soils showing considerable laterization, and also grey-brown soils exhibiting some degree of podzolization. The former resemble the tropical red soils, but have much more shallow B horizons (Fig. 235) and often a grey surface layer, but ploughing reveals the reddish colour. As with the savana soils, exhaustion is rapid because of the scanty organic material and the leaching of soluble minerals in the upper horizons. The grey-brown soils are actually more common in cool temperate regions and are discussed on p. 329. A group of intrazonal soils in the eastern marginal areas are the black earths; some are like the chernozems (see p. 330), whilst others have limestone parents, as in the cotton belt of North America.

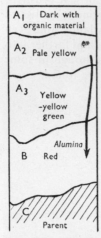

FIG. 235.—Profile of Warm Temperate (Subtropical) Red and Yellow Soil (South-east U.S.A.)

The west marginal areas have, in general, less leached soils on account of the summer drought. Most of them are pedocals with fairly dark surface layers which pass down into lighter horizons containing calcium compounds. Two important zonal soils are the brown (steppe) soil and the chestnut soil (see p. 330). A red intrazonal soil is developed on a parent of limestone, the latter being a common rock around the Mediterranean Sea basin. This soil is mainly a red clay, left as a residue of iron oxides by the solution of several hundred feet of limestone. It is known as 'terra rossa', although this term is applied to red soils in many parts of the world. The grey-brown soils referred to above are also found in western areas with more plentiful rainfall.

The semi-arid and desert areas between the east and west margins have brown pedocals and desert soils of the type described on p. 295.

CHAPTER XIX

COOL TEMPERATE CLIMATES

Distribution and Major Controls

The cool temperate regions lie on the poleward side of latitude 40° in each hemisphere. The major areas are in North America and Eurasia, where there is a marked contrast between the interiors and the margins (Fig. 236). In the southern world the areas are small and do not permit of the threefold division mentioned on p. 296.

The west margins are in the direct path of the westerly winds which blow at all seasons, but depressions are so common that the climate may be regarded as a sequence of weather which shows some sort of seasonal rhythm. The east margins are much influenced by the seasonal pressure distributions of the interiors, but again in the east depressions play their rôle, whilst in eastern Asia there are monsoonal controls of some importance.

Temperature

Now as well as one to five months with mean temperatures below 43° F., there are much wider ranges of temperature to distinguish these areas from warm temperate regions. In fact, the very wide ranges experienced in the interiors make the term 'temperate' quite inappropriate. Nevertheless, the western margins have very equable climates, since there is a fairly regular importation of warm air of tropical origin, particularly in the winter. The oceans adjacent to these western margins in the north have large positive anomalies of temperature in January. In effect this means that the eastern parts of the North Pacific and Atlantic Oceans are at least 20° F. warmer in January than would be expected for such a high latitude. The course of the January reduced mean isotherm of 32° F. demonstrates this quite well, as it runs from the Pacific shores of Canada south through some twelve degrees of latitude to St. Louis. A similar course may be noted from Norway to the Black Sea, thus enclosing an extensive zone of the Atlantic margins, including the British Isles, known as the gulf of winter warmth.

Increasing continentality towards the interior of each northern land mass is shown by the mean annual ranges of temperature at successive stations lying from west to east in approximately the same latitude.

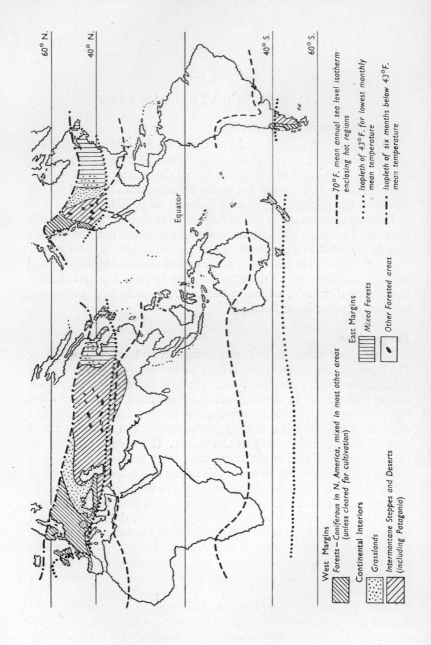

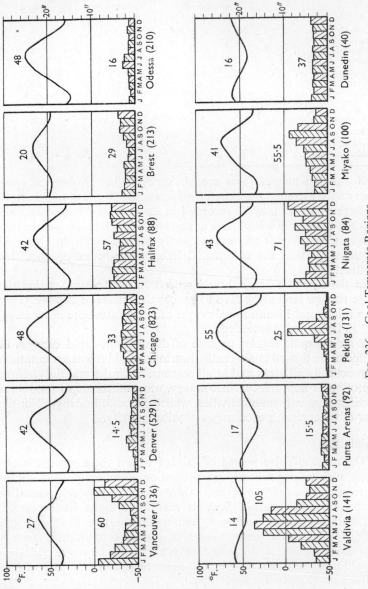

FIG. 236.—Cool Temperate Regions

	January Mean Temperature (° F.)	July Mean Temperature (° F.)	Mean Annual Range (° F.)
Eurasia:			
Valentia, 52° N. . .	44	59	15
Berlin, 53° N. . .	30	66	36
Warsaw, 52° N. . .	26	66	40
Kiev, 50° N. . . .	21	67	46
Vladivostok, 43° N. .	7	69 (Aug.)	62
North America:			
Portland (Oregon), 46° N.	39	67	28
Denver, 40° N. . .	30	72	42
Chicago, 42° N. . .	26	74	48
Boston, 42° N. . .	27	72	45

Hot-month temperatures are obviously higher in interior areas than on the margins, and extreme mean annual ranges of temperature as large as 80° F. may be recorded in the interiors. East coasts are markedly colder in winter than at corresponding latitudes on western shores. Cold currents offshore are partly responsible for this, but cold air moving from the continental interiors helps to produce this contrast.

In the southern hemisphere large ocean surfaces separating the quite narrow land areas help to reduce the effects of continentality to a minimum. Mean annual ranges of temperature are never large, 20° F. being rarely exceeded.

With the quite regular passage of depressions, rapid changes of temperature related to alternating warm and cold fronts are common in practically all areas and at all seasons. Such thermal fluctuations are of some importance, since muggy, warm air tends to produce a decline in most human activities, whilst the cold, bracing air in the rear of a depression encourages physical activity.

Rainfall

The most effective source of rainfall within these regions is un-questionably the westerly wind circulation, although it must always be remembered that these winds are more often than not related to depressions. In general, the west margins are the wettest areas, particularly on coasts backed by high mountains (see Vancouver and Valdivia), but on the leeward slopes of these mountains there is much less rain (see Punta Arenas and Denver). The interiors have much less rain, but in the two northern continents there is an increase of rain as the east margins are approached.

Frontal, relief, and convectional rainfall prevails in the various areas and at different seasons, so that distinct régimes may be recognized;

1. Both west and east margins have well-distributed rainfall which is mainly frontal and relief in origin (see Brest and Halifax).

2. The continental interiors have convectional rains, often from thunderstorms in summer, coinciding with the greatest heat (see Odessa). In spring depressions may bring additional rainfall to the western parts of the interiors (see Denver).

3. Eastern Asia has a marked summer maximum of monsoonal rainfall which is both frontal and relief (see Peking).

Transitions between the well-distributed rainfall of the western areas and the summer rains of the continental interiors are commonly found. For instance, Paris receives more rain in summer than at other seasons, and yet in the whole year there is less rain here than at Brest in the west of France. In all cool temperate areas snowfalls are usual in winter, but rarely does snow lie for long in the western margins except on very high ground.

Vegetation

It must be pointed out that in many of these areas the natural vegetation has disappeared where land has been cleared for farming. In the western margins with mild winters there is only a short period when temperatures are low enough to stop plant growth, whereas in the interiors growth is arrested for several months.

Whilst it would be expected that the west margins with their abundant rainfall should support forest, the salt spray blown by the prevailing winds clogs the leaf pores, so that trees are restricted in coastal areas. It is the transitional areas of mid-England, eastern France, and much of Germany which carry, under natural conditions, broad-leaved deciduous forests of ash, oak, and beech in particular as well as birch (Fig. 204). The leaf fall marks a pause in growth in the winter period. In the western margins of the Americas there are coniferous forests, and in southern Chile there are Antarctic beech trees. It is only at high levels or on poor soils that conifers are found extensively in cool temperate areas of Europe. In the Scilly Isles and the Kerry coastlands of Eire, subtropical evergreens of 'Mediterranean' type flourish out of doors, but, from the course of the 43° F. mean isotherm for the coldest month, these remote western areas have, strictly speaking, a warm temperate climate.

The continental interiors, with light spring and summer rains, carry steppe and prairie grasslands, and in some better-watered parts mixed coniferous and deciduous forests. The grasses are relatively tough and dry in comparison with the meadow grasses of the western margins.

The eastern margins of the northern continents with well-distributed rainfall have also mixed forests of deciduous and coniferous trees.

Nearly everywhere in these regions the forests are good lumbering areas, particularly in the mixed forests, where the choice of timber is great. In high demand are pine, spruce, hemlock, Douglas fir, elm, oak, and beech.

Northern Regions

Since there are significant differences between the regions of the two northern continents it is best to consider each as a whole. Nevertheless, the major controls of their cool temperate climates may be discussed together.

(*a*) *Winter.*—As shown schematically in Fig. 237 there are three main pressure distributions, high pressure centred over each of the

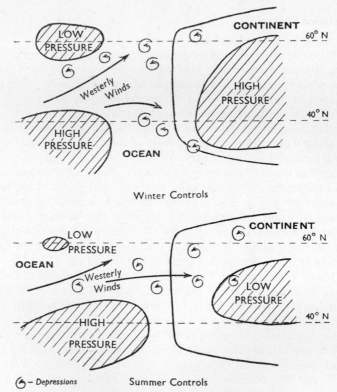

FIG. 237.—Seasonal Controls of Pressure in Cool Temperate Areas
(Northern Hemisphere)

continents, high pressure in the subtropical parts of the northern oceans, and low pressure which is centred over Iceland and the Aleutian Islands respectively. The straight arrows show the south-westerly winds, but these are essentially parts of the anticlockwise circulations of depressions as shown by the curved arrows. The courses taken by depressions are mainly north-east towards Scandinavia in Europe, although a few pass southwards through the Mediterranean Sea basin. These courses are largely determined by the buffer of high pressure over the continent. In North America the

→ January tracks
+++→ April "
- -→ July "
•••••◇ October "
(after van Bebber)

FIG. 238.—Tracks of Depressions in Europe

guidance and the obstruction of the western mountains cause a southerly circuit along the shores of the Gulf of Mexico to be the most favoured route, although some depressions pass directly across the continent in this season. All depressions in North America tend to issue from the Laurentian lowlands or New England into the Atlantic area (Fig. 239).

(b) *Summer.*—As shown by Fig. 237 the oceanic low-pressure area of winter has been relatively reduced by the development of major low-pressure zones over the land masses to east and west, whilst the subtropical oceanic high-pressure belt has moved north and extended eastwards towards the continent. Depressions are now much more prone to take courses which lead them into the heart of each continent, although in Europe there is still a strong tendency for the disturbances to follow tracks to the north-east.

It is possible to discriminate between the cyclonic tracks in the four seasons of the year. Thus in spring and to a less extent in autumn the

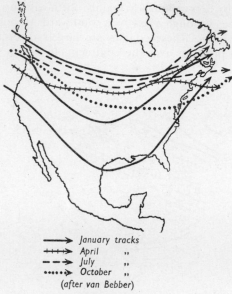

⟶	January tracks
+++⟶	April „
− −⟶	July „
•••••⟶	October „

(after van Bebber)

Fig. 239.—Tracks of Depressions in North America

Mediterranean corridor is frequented. However, it is undeniable that many depressions follow northern routes throughout the year in both continents (Figs. 238 and 239).

Eurasia

(a) Western Europe and the British Isles

Despite the vagaries of cyclonic weather there are fairly common weather patterns characterizing the various seasons. In winter the western margins often enjoy mild and humid weather when the continental high pressure is restricted to eastern Europe where intensely cold weather is experienced. Practically the whole of the seaboard of Western Europe, including Scandinavia, receives frontal rainfall from depressions and is under the influence of the gulf of winter warmth. Norway may have milder weather than places on the northern shores of the Mediterranean Sea. Occasionally in winter the continental high pressure extends considerably and covers the western margins. This extension of the high pressure may be very persistent, and with clear skies and therefore unimpeded nocturnal radiation severe frosts are common. An extension of the Azores high pressure towards Europe often brings fog and anticyclonic gloom as warm humid air is chilled by contact with the relatively cold land surfaces of the coastal regions. Radiation is reflected back from low stratus clouds, and frost is unusual. If the continental high pressure extends towards Britain then it may be quite cold, particularly in the south and east of England. Snowfall in Western Europe is often produced when a wedge of cold air protrudes westwards and forces tropical air upwards, thus giving abundant frontal rainfall which falls through the cold air and is converted to snow.

In spring, Western Europe is frequently the site of a large col through which depressions move rapidly from Britain towards the Mediterranean Sea basin when squally weather, at a succession of fronts, and typical spring showers are common. Many depressions pass south of Britain in this season, giving easterly winds over our islands. These winds are reinforced by the continental high pressure and the weather is cold and dry.

In summer, with the northward spread of the Azores high-pressure area and the development of low pressure in Europe, slow-moving anticyclones pass inland accompanied by intervening depressions. Sometimes the latter may be predominant, and the British summer is very wet. Rainfall is mainly convectional and often associated with troughs of low pressure and thunderstorms. A persistent anticyclone may give a heat-wave lasting for several weeks in England, but usually the settled weather of an anticyclone lasts for perhaps nine or ten days, when it is replaced by cyclonic weather. At this time of the year Eastern England has more rain than at any other season, thus showing the continental feature of a summer maximum of convectional rainfall.

Autumn is somewhat like summer, with anticyclonic weather which may last until late October. Night radiation is considerable, so that frosts and dense mists are not uncommon. Depressions give the western coastlands much frontal rainfall, which may be heavy, since warm, humid air is moving on to a land surface which is cooling steadily (see Brest and Valentia).

(b) Interior Zone—Eastern Europe and Inner Asia

As a whole in Europe the rainfall decreases towards the east, although an extensive region, reaching at least 1,000 miles from the Atlantic seaboard, has over twenty inches of rain in the year. Eastwards the significant change is from rain at all seasons to a markedly summer maximum of rain, but parts of the Russian steppes receive spring rain from depressions. Sheltered basins, such as the Alföld of Hungary, are drier than would be expected for so westerly a position, and carry areas of steppe and semi-desert. On the other hand inland seas increase the humidity of local winds, so that there may be quite large totals of rain, as at the southern shore of the Caspian Sea, where forests flourish. Eastwards from the Caspian regions, increasing continentality, combined with the physical nature of the intermontane basins, produces steppe and desert in inner Asia (Fig. 234). The observations already made on temperate deserts (p. 308) are relevant.

(c) Eastern Margins—North China and Japan

In this monsoon region intensely cold air from inner Asia covers North China, and also affects Japan, in winter. Peking has mean temperatures below 32° F. for three months, and much of northern Japan is similarly very cold (there is also a cold current of water moving towards Japan from the north). Depressions, moving out of China, draw air from the seas, which gives light snow in North China. Many of these depressions follow the two streams of the Kuro Siwo as it branches and passes along both sides of the islands of Japan. With outblowing winds from China, which have to cross the Sea of Japan and gain water vapour, and with these depressions, heavy winter precipitation of both rain and snow is experienced on the northern and western shores of the Japanese islands, whereas their south and east coasts are both milder and drier (compare Niigata and Miyako).

In summer onshore winds give a maximum of rainfall in North China and Japan, although in the latter area the north and west coasts are less rainy than those on the south and east. Additional rain may come in spring from depressions and in autumn from the northern fringes of typhoons whose active centres are well to the south. Mean temperatures in the hot months approach 80° F. in North China, but the oceanic influence in Japan prevents temperatures rising much above 75° F., and the hottest month may be delayed until August. Within so limited an extent of latitude (about fifteen degrees) Japan shows a remarkable range of vegetation passing from coniferous forest in the north, through deciduous forest to the southern areas, where subtropical plants such as bamboos and camphor trees flourish.

North America

The west marginal zone of this continent is very limited in comparison with that of Europe, and the intermontane basins and plateaus of the western cordillera carry steppe and desert to within 200 miles of the west coast. The eastern part of the continent has, however, a very extensive region, which receives adequate rainfall as far west as about 100° W., some 1,000 miles from the east coast.

(a) West Margins—Pacific Province

In winter the controls are as shown in Fig. 237 but the air accompanying the North Pacific Drift is not so warm nor as humid as that in the North Atlantic Ocean, so that January mean temperatures are some 5° F. lower than in corresponding latitudes on the European seaboard. The high mountains near the western coasts limit the area which enjoys relatively high winter temperatures, as well as the heavy frontal rainfall from depressions. Nevertheless, annual rainfall of over 100 inches is common near sea-level on the Canadian coast.

In summer with the northward shift of the high-pressure cell in the North Pacific Ocean, prevailing winds are north-westerly, but relatively few disturbances pass into the land, and there is only moderate rainfall, which decreases in Oregon as the 'Mediterranean' area of California is approached.

This coastal province has probably the densest coniferous forest in the world and includes Douglas firs (Fig. 205).

(b) Interior Areas—Basin and Ranges Province and the High Plains

This region shows some of the characteristics of inner Eurasia. The continental nature of the climate is shown by the large mean annual ranges of temperature, usually well over 40° F., with January means below 32° F. and July means of the order of 70° F. (see Denver). Depressions are responsible for much of the spring and summer rainfall, but the annual totals are only about 15 inches (in some parts less than 10 inches are received) which is in contrast to the western coastlands. Immediately east of the Rocky Mountains there is little winter precipitation, and the continentality is shown by the well-defined summer maximum of convectional rainfall. In summer there may be tornadoes which are the most intense form of temperate air swirl, usually associated with troughs of low pressure. As in the Russian steppes, so here spring rain from depressions is not uncommon.

The Basin and Ranges area carries sage brush and desert, but the high ranges have appreciable rainfall on their western slopes and are clothed with thin coniferous forest. The High Plains to the east have steppe and short prairie grassland, but some very arid areas have the bare rock and sandhills of the 'bad lands'.

(c) Eastern Areas (mainly east of 100° W.)

This area lies north of the cotton belt, extending to the St. Lawrence lowlands and the Ontario Peninsula. Its great width gives it both marine and continental characteristics. In winter depressions are limited, since high-pressure air restricts the majority to the southern circuit (Fig. 239). However, those which pass through the eastern states draw cold humid air off the Atlantic Ocean, and thus bring both rain and snow. In summer continental heat and low pressure produce convectional rainfall everywhere, and this is supplemented by frontal rain which, in the west, is delivered from tropical air drawn from the Gulf of Mexico, whilst in the east such rain is derived from Atlantic sources.

Winter temperatures are not excessively low, since the water bodies of the Great Lakes (which are never completely frozen over), like the oceans, have a moderating influence. Winter isotherms show a northward bend over the Great Lakes region. Toronto is warmer than Montreal in the winter months.

Chicago exemplifies the continental west of this great area, with a large mean annual range of temperature (48° F.) and a summer maximum of rainfall (33 inches per annum), whilst Halifax is typical of the marine east, having a range of 41° F. and well-distributed precipitation (57 inches per annum).

In winter cold waves of polar continental air or polar maritime air, the latter bringing heavy snowfalls, are associated with depressions. Hot waves are produced in summer, when tropical air in the warm sectors of almost stationary depressions raises temperatures to over 100° F. on several days (night temperatures rarely fall below 75° F.).

The natural vegetation of this eastern region consists mainly of mixed deciduous and coniferous forest, but today there are few extensive forested areas apart from the uncultivated uplands such as are found in the Appalachian Mountains. The forests pass westwards into tall prairie grasslands.

Southern Regions

The areas of Southern Chile (with Patagonia), Tasmania, and New Zealand have several features in common. There are no distinct eastern margins, nor do interior areas occur. All of these areas are close to extensive oceanic surfaces, so that the mean annual ranges of temperatures are usually less than 20° F., and only the southernmost parts of South America and New Zealand are high enough in latitude to have mean winter temperatures below 40° F. All lie in the Roaring Forties with westerly winds and disturbances which, on meeting the elevated terrain that lies on all the western shores of these areas (apart from North Island, New Zealand), give excessively heavy annual rainfall; many coastal stations have over 100 inches (see Valdivia). Eastwards in the lee of the mountains the rainfall decreases, so that parts of Patagonia have less than ten inches and in both Tasmania and New Zealand there are areas with less than twenty inches of rain in the year.

South America

Southern Chile and Patagonia are extraordinarily contrasted, with dense forests of conifers and Antarctic beech trees as well as much swamp in the former, whilst in the latter there is steppe and even desert. However, in Patagonia the small rainfall is well distributed throughout the year, and in some valleys, fed by Andean streams, woodlands appear, whilst an acacia scrub, called espinal, is found in the Rio Negro district.

New Zealand

These islands enjoy the oceanic moderation of the summer heat, and some places have hot-month temperatures which are considerably lower than places in corresponding northern latitudes. Most of New Zealand appears on the map in Fig. 236 as a warm temperate region. However, whilst the south is strictly cool temperate, in many ways the whole region is of this type. Whilst westerly winds prevail over most of the country, the northern areas of North Island receive trade winds with rain from the south-east in summer. Rainfall is well distributed during the year, since there is a constant succession of depressions (see Dunedin), and yet there is a very high incidence of sunshine in these islands. In South Island the mean annual isohyets trend parallel to one another and to the longer axis of the island, and there is a decrease of rainfall from west to east. North Island does not exhibit this pattern of rainfall distribution, as its relief is more complex. New Zealand has temperate rain forests, with Antarctic beeches and other trees, such as the kauri pine in the north, whilst grass with woodland flourishes in the drier parts of the country.

The soils of the cool temperate regions are discussed with those of cold temperate areas (p. 329).

CHAPTER XX

COLD CLIMATES—SOILS OF COOL TEMPERATE AND COLD REGIONS

Types and Distribution of Cold Climates

Polewards from cool temperate regions there may be distinguished three types of climate, two of which experience mean temperatures that permit only a very short growing season for plants. The three types of climate are;

(i) *Cold Temperate Climate.*—With at least six months having mean monthly temperatures below 43° F.

(ii) *Tundra Climate.*—With a warm-month temperature of not more than 50° F.

(iii) *Polar Climate.*—With perpetual frost and without plant life.

(a) Cold Temperate Climates

These climates are found exclusively in the northern hemisphere (Fig. 240). Lying in the westerly wind belt, the regions have climates which are much like those of cool temperate areas. West margins are very restricted in both the northern continents, since western coasts are backed by high mountains. Very wide interior zones have the usual characteristics of continentality. Eastern margins are less clearly differentiated but in Siberia there is a monsoonal control.

Temperature

In the narrow west margins mean annual ranges are small, usually between 20° F. and 30° F. In winter the gulf of winter warmth is an important factor in maintaining the whole Norwegian coast and much of its North American counterpart free from ice. The course of the mean January 32° F. isotherm lies slightly inland along the west coast of each continent. Highland restricts the warming influence of the ocean, and the heads of the larger fjords of these coasts are often icebound, having January mean temperatures some 10° F. lower than at their sea ends. In the summer the tempering effect of the ocean prevents the west coasts from having very high temperatures (Bergen —hot-month temperature of 58° F.); in addition there are frequent fogs which have a cooling influence.

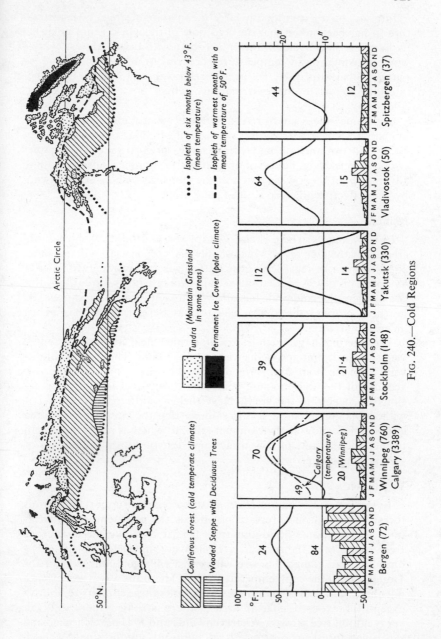

•••• Isopleth of six months below 43° F. (mean temperature)

– – – Isopleth of warmest month with a mean temperature of 50° F.

Coniferous Forest (cold temperate climate)

Wooded Steppe with Deciduous Trees

Tundra (Mountain Grassland in some areas)

Permanent Ice Cover (polar climate)

Arctic Circle

50° N.

Bergen (72)   Winnipeg (760)   Stockholm (148)   Yakutsk (330)   Vladivostok (50)   Spitzbergen (37)
                Calgary (3389)

Calgary (temperature)   Winnipeg

Fig. 240.—Cold Regions

Interiors have exceptionally large mean annual ranges of temperature. It has been noted that the cold pole of the world is not polar in its location, since it lies within north-eastern Siberia. Almost everywhere in winter mean temperatures descend to at least 0° F., and in parts of Siberia to —48° F., as at Verkhoyansk. Cold air from the interior spreads to the east margins, which are only slightly less cold.

In summer these interior regions have mean temperatures in the hot month which nearly always exceed 65° F. and may reach above 70° F. Spring and autumn are virtually non-existent. Within the Arctic Circle long hours of insolation in summer offset the disadvantages of the low elevation of the sun's rays, and insolation can by very effective on steep south-facing slopes.

Rainfall

In west margins there is considerable frontal rainfall which is increased by the high relief. Much snow falls, even at sea-level, in winter and this must be regarded as the wettest season, although summer is by no means dry. Annual totals of rain decrease northwards along the coast where the colder air has a lower capacity for water vapour. In northern Norway the total is half that at Bergen, which has eighty inches per annum.

The interiors have much less rainfall than the coasts and there is a very sudden decline across the western mountains. These areas enjoy a very large freedom from rain or snow-bearing winds in winter on account of the high-pressure control, although occasionally depressions do penetrate the areas. Nevertheless, although winter is not without its precipitation, it is in summer that the interiors receive the largest proportion of their annual rainfall, which is rarely more than twenty inches, and of this total some ten or fifteen inches are of summer incidence.

Vegetation

Both west coasts and interiors of these regions support coniferous forests which, in fact, replace polewards the cool temperate deciduous forests. Some deciduous trees persist but, like the coniferous trees, other than the spruces, are relatively undeveloped (Fig. 241), and none can compare in size with either the deciduous trees or the Douglas fir of the cool temperate forests. The short warm season with adequate rainfall even in the interiors permits these conifers to flourish except on exposed coastal areas, where strong winds and salt spray inhibit tree growth. Wherever winds and fires have checked and destroyed forest areas or where rainfall totals are too small, wooded steppe replaces forest, although there is an unbroken belt of forest across both Canada and Eurasia (Fig. 111).

Regional Types

North America

(a) *West Coast—Alaska and British Columbia.*—The maritime conditions which have been described above are applicable to this region. Mountains rising to well over 15,000 feet give excessively high annual totals of rain and snow, over 100 inches at some stations, although, in

E.N.A.

FIG. 241.—Coniferous Forest, near Maals River, Norway
In the foreground are some deciduous trees. In the uplands may be seen corries.

general, there is a decrease from British Columbia in a northward direction.

(b) *Intermontane Zone of the Cordilleran Region.*—Behind the forested coastal mountains the Alaskan interior has less than twenty inches of rain per annum. Much of the area is very high and winters are very cold. The predominant vegetation is tundra.

(c) *Canada, east of the Cordilleran Region.*—Most of this area may be regarded climatically as a continental region, but winter high pressure is less rigid in its control than is the case in Eurasia. Depressions bring not infrequent blizzards with their intensely cold, humid air of polar origin, which makes the winter chill very irksome. Immediately east of the Rocky Mountains rapid increases of temperature are experienced at the time when the chinook wind blows (see p. 223). Although this wind is not regular in its occurrence, it has a frequency which appears to affect the mean winter temperatures of West Albertan towns that, despite their being more northerly and at a greater elevation, are actually rather warmer than Winnipeg (see Winnipeg graph with Calgary temperature curve).

Towards the east of Canada, frequent depressions issuing by the Laurentian region into the North Atlantic Ocean have, it is thought, a moderating effect on the winter chill, and also the water masses of the Great Lakes, as referred to on p. 322, play their part in reducing the severity of the season. Both Labrador and Newfoundland are affected by a cold current which in summer keeps the mean temperatures below 60° F. Fog is also common on these coasts, particularly in winter.

Immediately east of the Rocky Mountains, the rainfall is, in general, less than twenty inches for the year. The western part of this zone has a typical continental summer maximum of rain that passes eastwards into a region of well-distributed precipitation with somewhat greater annual totals nearer the Laurentian lowlands and in Newfoundland. An annual total of 20·2 inches at Winnipeg increases to 54·6 inches at St. John's, in Newfoundland.

Eurasia

(a) *North-west Europe.*—The contrast between Norway and Sweden with Finland may be regarded as an epitome of the contrast between maritime and continental regions in high latitudes. Whereas the Norwegian coast is almost completely ice-free, the Gulf of Bothnia is frozen over for some six months of the winter and spring. Although continentality is shown by the summer maximum of convectional rainfall in Sweden, the summer temperatures are lower than those experienced in Siberia (see Stockholm and Yakutsk).

The rainfall on the Norwegian coast is somewhat lighter than in western Canada and Alaska, since the mountains are not more than 6,000 feet in height. In winter Norway may suffer from very cold easterly winds derived from the continental high pressure, although mild Atlantic weather is more common. These cold winds create advective fog as they pass out over the warm water. Depressions, with west winds of gale force, affect the Norwegian coast on an average once every week in winter, and give a winter maximum of rainfall and snow. There is no föhn effect, as in Canada, to the east of the Scandinavian mountains.

(b) *Interior Eurasia*.—Here the continentality is most pronounced (see Yakutsk), but despite the intense winter cold the air is dry, and only rarely do depressions bring humid polar air with the wind known as the buran. The summer rainfall is sufficient, with temperatures of over 60° F., to support coniferous forest.

(c) *Eastern Siberia*.—The whole Siberian coast as far south as Vladivostok is frozen in winter since it is under continental influences, and in northern sections temperatures fall almost to the level of those at Verkhoyansk. In summer, inblowing monsoonal winds directed to the continental low pressure give coastal rainfall, sometimes in excess of twenty inches, which declines inland and fails at the eastern edges of the interior plateaus. These winds blow from cool seas and prevent summer temperatures even in the south from rising above 70° F. As in the interior zones, so here winter passes directly to summer and vice versa, so that there is neither spring nor autumn.

(b) Soils of Cool and Cold Temperate Regions

From the point of view of soils it is most satisfactory to discuss those of the cool and cold temperate regions together. In the wetter areas are the zonal podzols and the grey-brown podzolic soils. The podzol is the soil of the Russian coniferous forests, and is grey at the surface (Fig. 242). The needle leaves from the trees form a thin layer of raw humus, below which, to a depth of about two feet, is a strongly leached horizon, which is ash-grey in colour. Below this lies a brown-coloured zone enriched with iron and aluminium compounds from above. With the relatively short summers of these regions bacterial action is limited and the humus remains very raw. Soil water is therefore fairly acid, and is responsible for much of the eluviation by chemical solution in the upper horizons. The structure of the soil is poor, and consequently it is of little value unless improved by artificial means.

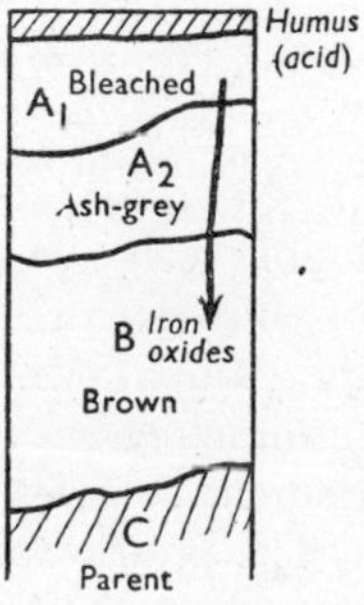

FIG. 242.—Profile of a Podzol

The grey-brown podzolic soils are much more widespread and of greater value than the podzols. They are common in the west and east margins where the rainfall is well distributed. Although leached the upper horizons are not bleached in colour as in the case of the true podzols. Brown hydroxides of iron remain near the surface and give, with the organic material, the typical grey-brown A horizon (Fig. 243). The lower horizons have less organic material and the colour is yellowish-brown. These soils are very productive in comparison with the podzols. Most northern areas that are covered with ground moraine have these podzolized soils.

Whilst podzolic soils are found in the fairly dry interiors, two other zonal soils are of importance, namely, the chernozems and the prairie soils. Chernozem, which is a pedocal, is the true black earth of the Russian steppes, but it is also found under prairie grasslands in other areas. Its profile (Fig. 244) shows a dark upper horizon (as deep as five feet) with abundant calcium compounds as well as less soluble alkaline salts which have been lifted by capillarity, since the rainfall is light and there is but slight leaching. The grassland cover ensures a high humus content. The lower horizons are not very clearly separated from the dark higher ones which are good clays of excellent structure because of the lime content. Chernozems are

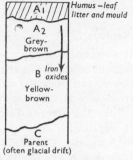

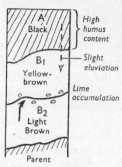

FIG. 243.—Profile of a   FIG. 244.—Profile of a
Grey-brown Podzolic Soil   Chernozem

highly retentive of water, and will give heavy crops without the need of enrichment by artificial manures. In the savana regions and the Pampas similar soils are developed, but the higher temperatures in these regions induce a more rapid decomposition of the humus, so that the soils are less rich in organic material.

The prairie soils are found in close proximity to the chernozems, but where there is rather more rainfall. While less rich in lime, the surface horizons show no podzolization. The A horizon is dark with humus, but below there is a grey-coloured eluviated layer (Fig. 245). Both chernozems and prairie soils are of high value and the temperate wheat and corn belts are closely related to them.

In the more arid interiors brown steppe and chestnut soils are developed. These pedocals have brown rather than black humus and a zone of lime accumulation which lies very high (Fig. 246). Parts of the intermontane plateaus of the northern continents have these soils. Crop production is not impossible on the chestnut soils, but usually they are left with a cover of natural grass for grazing purposes. Elsewhere grey desert soils are found in these remote interior regions.

Soils on loess are similar to steppe soils, although loessic soils must be regarded as almost intrazonal in character. Loess is formed as a rock by æolian deposition (see p. 125), but the finely comminuted rock material, the raw mineral content of soil, is readily changed into a loessic soil by the humus derived from the grass cover, and in most cases the soil is very fertile indeed. Both chernozems and prairie soils may have a loess parent.

Intrazonal soils in these temperate areas include rendzinas on chalk and limestone, and there are also peat soils. Peat, like loess, is a rock, but it contains the raw organic material of a soil which is readily converted into a true soil. Fen peat soils form in marshes whose water

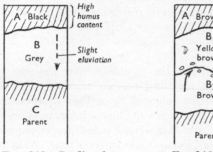

FIG. 245.—Profile of a   FIG. 246.—Profile of a Brown
      Prairie Soil   Steppe (Chestnut) Soil

contains dissolved alkaline minerals. Without such alkaline waters the peat soil is acid, although this type of peat soil may be formed when vegetable matter decays in very wet ground, where the exclusion of oxygen results in the production of much acid humus. These peat soils are sometimes referred to as bog soils and are, in fact, reasonably common in the river plains of most regions.

Tundra soils are discussed in detail on p. 333.

(c) Tundra and Polar Climates

Both the other types of cold climate may be classed as polar in the sense that they are largely under the control of the polar caps of high pressure. The distinction between tundra climate with its short growing season and the polar climate with no period of the year which is free from frost reflects the relative importance of the former. At best the tundra climate is one that must repel man, so that the geographical importance of these regions is slight.

The Arctic region consists of a central ocean basin that is almost completely landlocked, whereas the Antarctic area is a land mass centred on the South Pole and reaches at several places the Antarctic Circle. Surrounding this continent of Antarctica is a complete ring of ocean, so that tundra is almost exclusively located in the northern hemisphere.

The regions are not known well and there exist few long-period records of meteorological observations. The dominant feature on the ground is the outblowing winds from the polar caps of high pressure, and these are easterly. There is clear evidence that this easterly circulation is replaced at no great height above ground-level by a westerly one. This westerly drift is shown by clouds and by smoke from Mt. Erebus, the volcano near the South Pole. The high-pressure cap in each hemisphere covers an area which coincides fairly closely with the extent of the pack ice. Around these caps of high pressure move families of depressions in an easterly direction; these disturbances are particularly common in the Southern Ocean.

Temperature

In these very high latitudes the insolation over the year is quite abnormal, since there are very long periods when the sun's rays are cut off completely, which are balanced by periods when insolation is not interrupted for more than a few nocturnal hours in every twenty-four. Both at the Arctic and Antarctic Circles there is one complete day with sun and another without sun in each year. With increasing latitude, within these circles, the number of complete days in the year with and without sun increases to a maximum at each pole, which has a six-month day and a six-month night in each year. Since temperature may be regarded as the result of a balance between heat received by insolation and heat lost by radiation, the significant dates in the year are not the solstices but the equinoxes, because it is after the equinoxes that the long night, or day, commences near the poles. It is usual to expect the lowest mean temperatures after the solstice corresponding to the winter period for the hemisphere concerned, but such temperatures may be delayed in these regions until the following equinox (see Spitzbergen). Hot-month temperatures are similarly postponed.

A more important result is that, despite the long hours of insolation in the so-called summer, the angle of the noonday sun at any period is so low that the heat received is very slight. Furthermore, much of this heat is used in melting the ice rather than in heating the ground and the air lying above. The tundra zones are only slightly more fortunate than other polar areas for, although maximum temperatures may on occasions soar to 60° F. and 70° F., the majority of places have rarely more than three months with mean temperatures above 32° F. As would be expected, situations near the sea are the most favoured. Mean annual ranges of temperature in both tundra and polar climates are large, but the amplitude never reaches that of stations in eastern Siberia. The Arctic basin is ice-covered but the water below has a temperature of about 29° F. and this prevents the minima at the surface from being very low.

Precipitation

Under the prevailing high pressure of the polar caps the chance of precipitation is very limited and, although snow falls, the air is so dry that usually there is very little. At the same time there is probably even less chance of a great loss of moisture, because the melting of snow and ice is slight. Apart from coastal and marginal areas, where depressions and blizzards may bring heavier falls of snow, rarely do annual totals exceed the equivalent of ten inches of rain.

Vegetation and Soils

The tundra regions have, with the annual release from frost, a chance of supporting plant life, but the plants are adapted to the very limited period of growth and to the peculiar soils. The life-cycle of plants must be accomplished in as short a period as three months. Growth is undoubtedly encouraged by the long hours of sunshine,

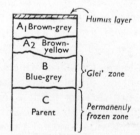

FIG. 247.—Profile of a Tundra Soil

but so handicapped are the plants that the normal process of seeding is largely abandoned and in its place there is vegetative reproduction.

After the mean temperatures have risen above 32° F. the soil is unfortunately of little worth, since it is clogged with water, and has a glei horizon (see p. 263). The soil is ice-cold and, as at no great depth it is permanently frozen, its drainage is largely impeded. Tundra soil is essentially anaerobic, and in it organic processes are inhibited. Little humus is formed; it remains raw and is therefore very acid (Fig. 247). Plants are compelled to behave in such a manner that they resemble the desert plants in that the frozen subsoil compels the root systems of the plants to spread extensively in the upper layers of the soil. Dwarf plants, including small trees of willow, alder, and birch, may succeed on better-drained slopes, but they have no great vertical development for quite slight winds will produce too rapid a

Dept. of Mines and Resources, Canada

Fig. 248.—Tundra in Summer, North-west Canada

The light patches are probably pools of water lying amongst the grass with low bushes, shrubs, and moss.

transpiration for their well-being. On slopes which are sunny and well drained, plants such as Iceland poppies, the saxifrages, and other familiar members of our British rock gardens may flourish. Elsewhere lichens and mosses, together with some coarse meadow grasses, are to be found, and these provide food for the reindeer and other animals of the regions (Fig. 248).

Part VI

OCEANOGRAPHY

CHAPTER XXI

OCEANOGRAPHY

Physical Features of the Ocean Basins

The oceans with their marginal seas occupy a very high proportion of the global surface. Some 70 per cent. of the world's surface is ocean and 30 per cent. is land. Despite this greater surface area of the oceans, our study must be very limited in comparison with that of the land surfaces and the atmosphere. Comparatively little is known about the oceans, and our chief aim will be to examine fairly closely such aspects of the oceans as have a bearing on meteorology and climatology. Nevertheless, the oceans possess certain fairly well-known features of submarine relief.

There are five oceans and they are all linked with one another. In the north is the Arctic Ocean occupying a basin surrounded by land masses and shallow seas. It is connected through the narrow Bering Strait to the North Pacific Ocean and by various water channels including, in particular, that of the Norwegian Sea leading to the North Atlantic Ocean. Both the Pacific and Atlantic Oceans extend through a considerable range of latitude which covers the northern and southern hemispheres and, as we shall see, they can be subdivided into several distinct sections. A fourth ocean, namely, the Indian Ocean, is less extensive in the northern hemisphere, but is otherwise not dissimilar to the Atlantic and Pacific Oceans, and in common with them it merges southwards into the Southern Ocean which surrounds the continent of Antarctica. All the oceans tend to be basin-like, with the possible exception of the last-named. In Fig. 1, p. 3, it will be seen that they separate the continental blocks of sial and that they are floored with sima, apart from the Atlantic Ocean, which is a depression in the sial layer.

All the ocean basins have certain common features of relief, although the relative importance of these features varies considerably. From the land margins the depth increases rather slowly in a zone known as the continental shelf, whose outer margin is marked by a great increase in the steepness of the sea floor until depths of 2,000–3,000 fathoms are reached (Fig. 249). At these depths is the most extensive feature of all oceans, namely, the abyssal floor or the oceanic plain. The link between the continental shelf and the floor is known as the continental slope. The floors of the oceans are remarkably level over very great areas, but they are interrupted by certain very marked deeps with very steep sides. The bottom of these ocean deeps may be five or six miles below the ocean surface. On the other hand the floors possess distinct ridges and broader areas known as submarine plateaus, covered by water which is often less than 2,000 fathoms in depth.

The continental shelf differs from the deeper parts of the oceans in several ways and one of these, which will be discussed more fully on p. 341, is related to the composition of the ocean deposits on the shelf as compared with those on the floors. This shelf varies very much in width as will be seen from atlas maps with bathymetric contours, which show 100-fathom depths. Although a depth of 100 fathoms is a fairly common one for the outer edge of the shelf, it is by no means unusual for the significant change in the slope of the sea

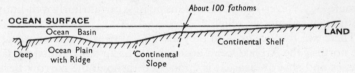

FIG. 249.—The Floor of the Ocean

Note.—In general, slopes are quite gentle (not to scale).

floor to occur at greater depths. What is so characteristic of the shelf is the relative plunge into the deeper parts of the oceans which occurs at its outer edge. The slope of the shelf has an average gradient of about 1°, whereas that of the continental slope is usually of the order of 5°. It is obvious that diagrams such as the hypsographic curve (Fig. 2) exaggerate these slopes very much, and it is equally clear that in relation to their widths most oceans are very shallow basins. The origin of this shelf is still debated. It would appear from Fig. 5 that the shelf is a submarine extension of a sial block. On this ledge there have accumulated land-derived deposits, and it has been argued that the shelf is largely, if not exclusively, a constructed feature. Depth surveys show that the shelves have submarine canyons crossing their surfaces which seem to mark the extensions of river courses from the land. The deeps which relieve the general monotony of level in the ocean plains are more often than not found near the margins of the basins and in close relationship with younger (Alpine) fold mountain systems, so that they are probably tectonic in origin.

The Atlantic Ocean is characterized in the northern hemisphere by quite wide continental shelves, but to the south of the Equator this is not so, apart from the shelf off Patagonia. In its most northerly parts the ocean is very shallow, and it is possible to regard the shelf as trans-oceanic, with the Norwegian Sea occupying a separate basin. South from this east–west shelf there extends an S-shaped ridge, which rises from the floor to levels of less than 2,000 fathoms depth. This ridge is almost parallel to the sinuous eastern and western coastlines of the Americas and of Europe and Africa respectively. Rising up from this ridge are several islands, including the Azores. The occurrence of this ridge is of interest, and supporters of the theory of continental drift argue that it represents the sialic material which was left when an earlier continent split into two parts to give the Americas and Eurasia with Africa. The ridge divides the ocean into two parts, and each of these is made up of two basins linked by shallow water along the equator. Marginal seas to the Atlantic Ocean fall into two groups. In the first group are the shallow seas, lying on the continental shelf, such as Hudson Bay with its allied straits and the North Sea with the Baltic Sea. The second group includes the Caribbean Sea and the Mediterranean Sea, which are very different from the seas of the first group, in that they contain several depressions of which the Blake Deep off Puerto Rico at over 4,500 fathoms may be mentioned.

The Pacific Ocean differs considerably in its details from the Atlantic basin. Whereas the latter tends to be of almost uniform width from east to west (it is least wide near the Equator), the Pacific Ocean is broadest in equatorial latitudes and narrows northwards. Southwards from the Equator it remains quite wide. The continental shelves off the various coasts of this ocean are, in general, quite narrow, but this is not so true in the Australasian parts of the basin, where a fairly wide shelf surrounds and links the East Indies with Australia and New Zealand, despite the occurrence of somewhat deeper areas here and there. This relatively shallow sea floor is joined southwards with Antarctica, and eastwards it has an extension in the form of a ridge which stretches almost to the narrow shelf off South America. The numerous island groups of Micronesia and Polynesia rise from a fairly shallow plateau in the centre of the ocean. Practically the whole of the ocean lying in the northern hemisphere is of a different nature from the foregoing areas. Here the floor is relatively deep throughout, and possesses some of the most notable deeps in all the oceans. It is believed that these great deeps are connected with tectonic disturbances which have played a great part in the development of the Pacific basin. Some of these deeps lie off the island arcs of Japan (Tuscarora Deep) and off the Philippine Islands (Mindanao Deep, about 5,500 fathoms), whilst others are apparently in close association with the several marginal seas off South China, Japan, and

Siberia, which resemble the Caribbean Sea in having quite deep waters. One other zone of the Pacific Ocean, with great deeps, lies in its southern part to the east of New Zealand and stretches towards the Fiji Islands.

The Indian Ocean needs little detailed description. The southern part is relatively shallow, whilst a series of shallow zones is related to several island groups which stretch from Madagascar towards India. The eastern part is the deepest, and like the western part of the Pacific Ocean contains some exceptional deeps off Sumatra and Java, as well as near South-west Australia. In this distribution of deeps the Indian Ocean is the reverse of the Pacific Ocean, its northern part being fairly shallow.

The Arctic Ocean has an inner basin about 1,500 fathoms in depth, which is surrounded by shallow seas that link it to the North Pacific and the North Atlantic Ocean, although the latter is further separated from the Arctic basin at one point by the Norwegian Sea. This sea has one area of very shallow water lying over the Wyville-Thomson ridge, which is no deeper than 200–250 fathoms in parts.

The Southern Ocean is a remarkably shallow ocean surrounding Antarctica, and it links up the southern parts of the Pacific, Atlantic, and Indian Oceans. It is thought that a ridge exists between South America and Graham Land, which supports the South Orkneys and other islands, whilst a broad plateau probably joins Australia with Antarctica.

Ocean Deposits

The floors of the oceans are covered with layers of sediments which belong to two classes. On the shelves and to a lesser extent on the continental slopes there are mainly terrigenous deposits (derived from the land surfaces), whilst in the deeper areas there are very fine deposits known as oozes, which are composed of the finely divided remnants of shells and skeletons of various sea organisms, together with fine volcanic dust and other material. The organic deposits are derived from various groups of sea creatures and marine plants, some being calcareous, whilst others are siliceous. Sea organisms may be divided into two groups according to their habitats. There are, for instance, the corals, mussels, and sea-weeds which are benthic, that is, they live, usually rooted, on the sea floor, and since they are limited to depths of not much more than 100 fathoms, they inhabit the waters of the continental shelves and add their quota to the terrigenous deposits. The other group of sea organisms are planktonic, in that they live in the surface waters of the seas and float about quite freely. This group includes pteropods and foraminifera, which have calcareous shells and structures, and radiolaria whose skeletal material is siliceous. In addition, there are diatoms or microscopic sea plants which have siliceous material in them. The diatoms are a source of food for most of the planktonic creatures. Since diatoms require light and this does not penetrate farther than about 650 feet in sea water, most planktonic life is restricted to depths of about 100 fathoms. Although most of these organisms are very minute, their numbers are countless, and they represent in all a much greater bulk than do the fishes and whales.

The terrigenous deposits have been briefly discussed in Chapter I, where it was shown that rivers carry out to sea particles of sand and clay together with carbonates in solution. The carbonates may be taken out of the sea water by various organisms in order to produce their shells, and on the other hand the carbonate may be deposited directly from sea water through temperature variations. Thus, the terrigenous deposits are sand and clay, which are laid on the shelf at varying distances from the land according to the size of their particles, together with some carbonate deposits derived from sea water in one way or another. There are also various coloured muds which are very finely divided clays. Their colour depends on chemical changes which produce different compounds of iron. Blue mud obtains its colour from ferrous sulphide, red mud from iron oxide, green mud from glauconite, a silicate of iron, which is formed in the presence of decaying organic material.

Large areas of the ocean floor beyond the continental shelves are covered with various oozes, and some muds and clays which consist largely of the minute remnants of the calcareous and siliceous organisms. Most oozes are greyish-white in colour. After death the shells of the organisms sink slowly downwards, and at the same time a considerable amount of solution of these shells occurs in sea water; they are small to begin with, rarely larger than a pin's head, but they become even thinner and smaller with solution. Calcareous remains are unusual below 3,500 fathoms, since their solution is more rapid than is the case with the siliceous material which may not be completely dissolved until depths of 5,000 fathoms are reached. Whilst the depth of the ocean floor determines to some extent the type of deposit that will accumulate in a given area, there is a further control, namely, that of temperature in the surface waters. For instance, diatoms, whilst occurring in all surface waters, are most common in the colder seas, and the most extensive deposits of diatomaceous ooze are on the floors of the Southern Ocean on the equatorward side of the terrigenous deposits which are derived from Antarctica. On the other hand, radiolaria flourish in warm tropical seas, and their siliceous oozes are found in the deeper parts of the Indian and the Pacific Oceans.

Below 5,000 fathoms most material of organic origin is absent and the commonest deposit is red clay, which is composed of fine volcanic ash, wind-borne dust from the land surfaces, minute remnants of the bones and teeth of large sea creatures such as whales and sharks, and even some terrigenous deposits that have been released in the open oceans by the melting of icebergs. This clay accumulates extremely slowly, and long-period decomposition produces much iron oxide which is responsible for its predominant red colour. It must not be thought that red clay is restricted to the great depths of the oceans, as it is a constituent of most oozes. Radiolarian ooze is, in fact, red clay with a high proportion of radiolarian remains, and the same is true of calcareous and other deposits.

Of the many organisms which produce calcareous oozes, two are outstanding, namely, the pteropods and Globigerina, the commonest of the foraminifera. The oozes of the former are restricted to fairly shallow depths, usually no more than 1,000 fathoms, and to tropical and subtropical seas where they are associated with the ridges and plateaus rising from the ocean plains. This pteropod ooze has a very limited distribution, mainly in quite small areas on the central ridge of the Atlantic Ocean, but other areas are found in the Mediterranean Sea and in parts of the Pacific Ocean off Australia. The Globigerina ooze has, on the other hand, a very wide distribution, particularly in the Atlantic and Indian Oceans, and to a less extent in the Pacific Ocean. Whilst foraminifera are capable of standing quite wide ranges of temperature in the surface waters of the ocean, they are usually found between 60° N. and 60° S. latitude, with the exception of the North Atlantic Ocean, where relatively warm surface water enables these deposits to form in latitudes as high as 70° N. This ooze is limited in depth by the complete solution of the calcium carbonate of the shells at about 2,500–3,000 fathoms, and towards these levels red clay becomes an increasingly important constituent and finally replaces the calcareous ooze in deeper waters. There are also certain zones of the oceans, around coral islands, where a distinct coral mud is deposited.

Temperature of Ocean Water

The temperature of the surface of the sea varies in much the same way as that of the land surfaces, since insolation is responsible for the varying quantities of heat which are received at different latitudes and in different seasons. However, there are significant differences between water and land surfaces which have been reviewed on pp. 163–4, and as water is a liquid, there is a distribution of the heat by mixing which is, of course, impossible with land areas. Surface temperatures of the oceans show a fairly considerable range during the year at most latitudes and in some zones, admittedly limited, this range may be as great as that which is experienced in certain continental interiors. Although ranges of 50° F. are on record, it is usual for the range to be of not more than 20° F. The surface temperatures vary with latitude, and mean annual temperatures of 80° F. and even higher are common at the surfaces of tropical seas, and there is a general decrease towards the poles where there are mean temperatures of slightly less than 32° F. Owing to sea water being saline, it may not freeze until temperatures of 29° F. or 28° F. are reached, the depression of the freezing temperature depending very much on the concentration of dissolved salts. There is not a constant decrease of surface temperature with increasing latitude, since drifting warm water from the tropical seas may move into high latitudes and give a local increase of temperature as in the North Atlantic Ocean, where there is, as has been inferred on p. 168, a positive anomaly of temperature in winter. On the other hand, upwellings of deeper and colder water may reduce locally the surface temperature of tropical and subtropical sea water.

The distribution of surface temperatures in the oceans during the year may be summarized as follows:

1. In tropical waters extreme temperatures may be as high as 90° F., or as low as 70° F., but the mean annual range is usually about 10° F.

2. In polar waters the lowest surface temperatures are experienced, approaching 32° F. (and may be lower, as explained above), and rarely does the temperature rise above 50° F. The mean annual range of surface temperature is usually of the same order as that for tropical waters, about 10° F.

3. The temperate seas which may be regarded as intermediate to the above zones have somewhat wider ranges of temperature. Adjacent to certain continental areas, as off eastern Canada and in a comparable area in the Sea of Japan, the mean annual range may be as large as 50° F. In general, however, it is between 20° F. and 40° F.

There is a fairly close connection between the temperature conditions of the surface waters and the sea organisms present. Thus, the high mean temperatures and the small ranges of the tropical waters encourage corals and many calcareous organisms, whereas in the cooler waters of higher latitudes the secretion of lime is much reduced and siliceous material is more commonly used for the structures of marine creatures. On the other hand, planktonic organisms are most abundant in temperate waters in fairly high latitudes, and this is reflected in the great fisheries of the subpolar seas.

Salinity of Ocean Water

The salinity of the oceans is produced by a large number of dissolved chemical compounds. We may regard the salt of the sea as the residue of solid material which is left after the evaporation of sea water. This residue can be weighed and can be expressed as a fraction of the whole weight of water before evaporation. Salinity is therefore defined as the weight in grammes of solid material left after the evaporation of 1,000 grammes of sea water. If the weight of solid material is 35 grammes (and it is usually very near this figure) the salinity would be shown as 35‰.[1] In the marginal seas of the oceans the salinity has a wide range of values. In the Baltic Sea fresh water entering from the surrounding lands reduces the salinity to 7‰, and it may fall in this sea as low as 2‰, but great evaporation combined with a very dry climate in the Red Sea region gives the water of that marginal sea a high salinity value of 41–42‰.[2]

Although a great number of salts can be identified in sea water, something like 80 per cent. of the salt is sodium chloride, whilst magnesium chloride, magnesium sulphate, and calcium sulphate together account for much of the remaining 20 per cent. In addition, there are very small quantities of many other salts which account for less than 1 per cent. of the total dissolved salts. Irrespective of the salinity value, whether high or low, the proportions of the different salts remain practically constant. This is explained by the thorough mixing which occurs throughout the waters of the several oceans. Estimates of salinity are therefore considered quite reliable when the content by weight of any one element in a sample of sea water is determined. The element used for this purpose is usually chlorine.

The origin of the various salts is somewhat problematical. Much of the salt content is derived from river water, but there is by no means an obvious relationship between the salts in river water and those in sea water. River water has a considerable content of dissolved carbonates, but these are relatively uncommon in sea water, having been abstracted for the formation of the shells of various marine organisms. On the other hand, chlorides, which have a low proportion in river water, become quite common in sea water largely because of the removal of carbonates in the sea, which thus increases the relative proportion of chlorides.

The salinity of open ocean waters varies within certain limits, although the value is rarely much above or below 35‰. A dilution of the salt water may be caused by the influx of water from large rivers,

[1] It is usual to express salinity in this way rather than as a percentage.
[2] Salinity of Dead Sea may be as great as 250‰.

by heavy rainfall, and in polar seas by the melt of ice, whilst a concentration of the salts is produced by great evaporation due to great
heat and strong winds. Dilution may be brought about by the transport of less saline waters into very saline waters and concentration
by the reverse movement.

Lines of equal salinity or isohalines may be drawn on maps from
salinity observations in the same way as isotherms may be traced. An
investigation of such maps shows that in rainy tropical and equatorial
zones the surface salinity values are low (below 35‰), whilst
immediately north and south of these zones in the areas affected by
the trade winds which cause much evaporation, the salinity rises to
36‰ and over. In temperate latitudes with fairly heavy rainfall at
sea, the salinity is reduced and decreases still more in a poleward
direction, so that in Arctic and Antarctic waters the value may be as
low as 33·5‰ and in some areas a value of 31‰ has been recorded.

In addition, sea water dissolves gases from the atmosphere, and
apart from nitrogen and oxygen there is a remarkably high content of
carbon dioxide. Bacterial action is also important in sea water, and
there is the fixation of nitrogen to form nitrogenous and ammonium
compounds. It is possible to regard these various dissolved substances
as playing a similar rôle in relation to the life within the ocean waters
as do the mineral and organic particles which affect the growth of
plants in the soil of the land.

Surface Circulations of the Oceans

The circulation of the waters in the oceans can be regarded in two
ways. There is a series of horizontal movements of water across the
surface of the oceans, but owing to the depth of the oceans there are
also both vertical and horizontal movements within the body of ocean
water. The complete circulation is extraordinarily complex, and can
be discussed but briefly in this book. Movements of water at any
level in the ocean body and in any direction depend upon the difference in the density of the water. The density depends upon the temperature and the salinity of the water, but movements at the surface
also depend upon the major wind flows which are controlled by
atmospheric pressure distributions. These surface movements are in
the form of currents and drifts. The former may be readily observed,
as they have speeds of 1–5 knots, and are of importance to navigation;
the latter are slow, almost imperceptible, movements of large masses
of water with a large surface area.

The Atlantic, Indian, and Pacific Oceans are sufficiently extensive
from north to south to have surface circulations which may be shown
by a generalized diagram for one ocean (Fig. 250). The most signifi-

cant features of this diagram are the two anticyclonic gyrals of currents that are clearly related to the subtropical oceanic cells of high pressure in each hemisphere. These two circulations approach one another in equatorial latitudes, so that westward-flowing currents are found immediately north and south of an eastward-flowing counter current. The two westward currents are thought to produce a piling

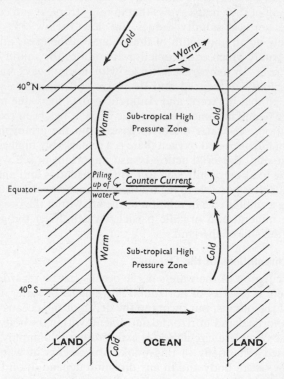

FIG. 250.—Generalized Surface Circulation in an Ocean

up of water on the western sides of the oceans, and some of this returns to make good the loss of surface water from the eastern side. There is in fact a surface gradient in the ocean with a gentle fall to the east. It is reasonable to assume that the North and South Equatorial Currents are induced by the surface drag of the trade winds of each hemisphere. On the western sides of the oceans these currents are deflected by the continental margins and turn poleward, moving into higher latitudes and eventually coming under the influence of the westerly winds when they become eastward-moving currents. In part, at least, these eastward currents are deflected towards the Equator on the eastern sides of the oceans. There is a shift of the latitudinal position of the currents in the equatorial zones, with the migration of the overhead sun during the year. This is most noticeable in the Atlantic Ocean. The diagram shows also that in high latitudes there are currents which move in an equatorward direction and are of polar origin. The diagram also distinguishes what are known as cold and warm currents. The former are currents which bring surface water that is relatively cold into areas of the ocean which are relatively warm, whereas the latter are opposite in character. Water from an area of the sea surface which has a low temperature will warm as it moves towards areas of warm surface water, but it will tend to have a temperature which is relatively low to that of the water that it enters and to that of the surrounding air, so it will have a cooling effect. The reverse is true of warm currents as they move water into cooler areas. A cold current may be caused by another factor, namely, the up-welling of deep water which is common on the eastern sides of the oceans in tropical and subtropical latitudes. Here the westward-moving trade winds in removing surface water from the eastern sides of the ocean induce deep water to rise and take the place of this dis-placed surface water. Although each of the three major oceans has a basic pattern of currents which is very like that shown in Fig. 250, there are certain distinctions characterizing each one. Whilst the southern parts of these oceans show a close similarity to one another, it is the parts lying north of the Equator that show the greatest diversity. The Indian Ocean is also somewhat exceptional, in that its northern limits are in low latitudes, whereas both the Pacific and Atlantic Oceans extend to very high northern latitudes. Figs. 251 and 252 show the essential points of resemblance and contrast between the circulations of these three oceans.

The two Equatorial Currents with their Counter-Equatorial Currents are different in the Atlantic and Pacific Oceans (Fig. 251). In the former water from both Equatorial Currents enters a westward-flowing current which passes mainly through the Caribbean Sea. Cape San Roque in South America seems to act as a deflecting point, so that whilst most of the South Equatorial Current turns southwards as the warm Brazilian Current, some of its water flows north-west-wards. Almost the exact opposite occurs in the western Pacific Ocean, where near the Philippine Islands the North Equatorial Current is partially deflected southwards, and enters the Equatorial Counter Current of that ocean. The current of the Atlantic Ocean which leaves the Caribbean Sea through Florida Strait becomes the warm Gulf Stream that can be traced as the West Wind Drift (that is, it loses the

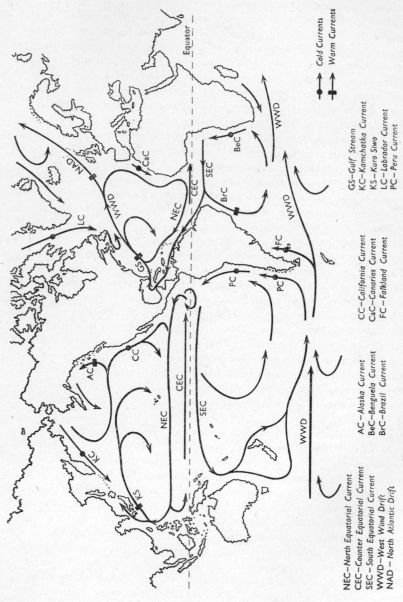

Cold Currents
Warm Currents

NEC—North Equatorial Current
CEC—Counter Equatorial Current
SEC—South Equatorial Current
WWD—West Wind Drift
NAD—North Atlantic Drift

AC—Alaska Current
BeC—Benguela Current
BrC—Brazil Current

CC—California Current
CaC—Canaries Current
FC—Falkland Current

GS—Gulf Stream
KC—Kamchatka Current
KS—Kuro Siwo
LC—Labrador Current
PC—Peru Current

FIG. 251.—Surface Circulations in the Atlantic and Pacific Oceans

Note.—Part of the North Equatorial Current in the Atlantic Ocean enters the Gulf of Mexico.

character of a well-defined current and broadens), and farther east it is known by yet another name, the North Atlantic Drift, which flows to the very northern limits of the ocean and enters the Norwegian Sea. It is true that quite large offshoots of the West Wind Drift occur; one swings southwards around the Sargasso Sea, and farther east part of this drift, usually known as the Canaries Current, turns south to join the North Equatorial stream. The counterpart of this quite exceptional current and drift in the North Pacific Ocean is much less extensive. The major part of the North Equatorial Current in the Pacific Ocean turns north near the Philippine Islands and proceeds as the Kuro-Siwo, a warm current, past the main islands of Japan and then eastwards as a drift, although it is highly doubtful if this reaches the coasts of North America. Like the West Wind Drift of the North Atlantic Ocean it sets southwards in a series of offshoots, the most easterly of which is in the neighbourhood of the Hawaiian Islands. Farther east there is a comparable current to the southerly drift on the east side of the northern Atlantic Ocean, which is known as the California Current, but this arises from an entirely separate gyral in the North Pacific Ocean to the south and east of the Aleutian Islands. Here water comes south and swings east, and then bifurcates into the northward-flowing Alaska Current which is thus relatively warm, and the southward-flowing California Current which is relatively cool. In the northern regions of both these oceans there are cold southward-flowing currents off Labrador and Greenland in the one ocean and off Kamchatka in the other.

South of the Equator both oceans merge with the Southern Ocean, and the anticyclonic gyral in each gives rise to a very regular pattern of currents. It will be observed that the southern margin of each gyral merges with the circumpolar West Wind Drift of water, and in part the waters of the southward-flowing and warm Brazil Current and of the similar current off Australia enter this circulation. The configuration of the coasts of Southern Chile and South-west Africa create a deflection of some of the West Wind Drift that moves north as a cold current along each of these coasts. Both the Benguela and Peru Currents are markedly cold, but investigations show that a complete account of them has yet to be given. For instance, the Peru Current, which has been very well studied, is not cold just because it flows from south to north, but owes its low surface temperature to much upwelling. It is not by any means uniform in its chilling effect, and recent investigations suggest very strongly that there are local minor swirls within the current at various places off the coasts of both Chile and Peru which cause warm water to prevail instead of cool or cold water. In all probability these are produced by influxes of relatively warm water from the ocean surface lying to the west of the main currents. Nevertheless, upwelling of water from the depths is regarded as the most important feature near the coast, although farther west there is what is known as the Peru Ocean Current, which consists largely of Antarctic water moving northwards. It has been suggested that the Falkland Current is caused in part by the offshore effect of the westerly winds, which induce cold water to flow northwards off Patagonia.

In the Indian Ocean the circulation north of the Equator is quite exceptional, since in this region the ocean surface is under the controls of the monsoonal circulation of the atmosphere. This means that whilst in the winter half of the year the wind direction is practically identical with the normal trade wind flow from the east, this direction is completely reversed in the summer half of the year. Consequently no North Equatorial Current flows in summer; in its place, and also apparently in the place of the counter-current, is an eastward-flowing current which may be referred to as the Monsoon Current (Fig. 252). In winter there is a westward-flowing current north of the Equator with a counter current between it and the South Equatorial Current of the Indian Ocean. In its southern part the Indian Ocean has a circulation which is comparable with that of the corresponding areas of the Pacific and Atlantic Oceans, with one slight difference, namely,

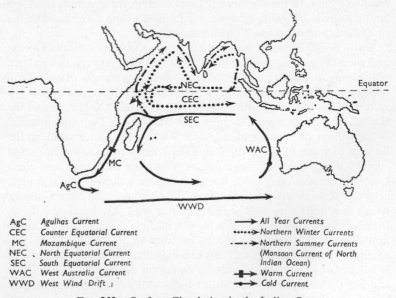

AgC	Agulhas Current	⟶ All Year Currents
CEC	Counter Equatorial Current	••••▷ Northern Winter Currents
MC	Mozambique Current	•—•▷ Northern Summer Currents
NEC	North Equatorial Current	(Monsoon Current of North
SEC	South Equatorial Current	Indian Ocean)
WAC	West Australia Current	⟶ Warm Current
WWD	West Wind Drift	⟶ Cold Current

FIG. 252.—Surface Circulation in the Indian Ocean

that the warm western current off Mozambique sends some water around the southern margins of Africa into the Atlantic Ocean. This flow is opposed to that of the main West Wind Drift and results in a series of local swirls or eddies. The cold current off western Australia is much less marked in its coolness than are the corresponding Peru and Benguela Currents.

Circulation within the Water of the Oceans

It has been shown that currents and drifts of surface water in the oceans can be linked closely with the frictional drag of winds across the surface, but it has been mentioned that these movements of water depend upon differences of density which can produce gradients in the surface waters, much as differences of atmospheric pressure will create pressure gradients and winds. The density of the surface water of an ocean can be both increased and decreased locally. Increase occurs when the surface waters are cooled, when there is great evaporation which increases the salinity, and where there is freezing of saline water. In the latter case the ice which forms does not contain all the soluble salts, and this means that the surrounding water which remains unfrozen has a greater concentration of salt and therefore a greater density. Decrease of density occurs when great heating causes the expansion of the surface water, also when great rainfall, or an influx of much fresh water from rivers or the melting of ice dilutes the water. When there is an increase in the density of sea water, as happens when saline tropical water moves into higher latitudes and cools, the water usually sinks to the level of the ocean which has the same density and there spreads out as a distinct layer. To compensate for this sinking of some surface water there must be a return of water to the surface. This may be in the vicinity of the sinking water, but it may occur in parts of the ocean very remote from where the water sinks.

In tropical latitudes the effect of great heating is to maintain a fairly low density in the surface waters, and even where high salinity values prevail there is little chance of the water sinking very deep. It is in high latitudes that water on chilling can become sufficiently dense to sink to the deeper parts of the ocean, and form what is known as bottom water and deep water. The most effective areas for this deep sinking are in certain parts of the North Atlantic Ocean and off Antarctica. In the former ocean, saline tropical water moving north with the Gulf Stream and its continuation, the North Atlantic Drift, becomes so very chilled in the seas between Greenland and Iceland that it sinks to very great depths. Off Antarctica the water not only becomes very dense through chilling, but its salinity is increased through the formation of much ice in this region. Much of this water finds its way down into the lowest levels of the South Atlantic Ocean (Fig. 253), and creeps at a very slow pace (it is highly improbable that it has a speed of more than one or two miles a day) northwards to the mid-latitudes of the North Atlantic Ocean. The temperature of this water is very slightly higher than the freezing temperature of sea water, that is, about 28° F. The densest water in the oceans is thought to lie on the floor of the Arctic Ocean, but it cannot move out of this basin, since it is hemmed in by the surrounding shelves of the shallow seas. There is little chance of water sinking to displace this bottom water in the Arctic Basin, since the surface waters of the Ocean have a very low salinity because of the great discharge of fresh water into it, particularly from the great rivers of Siberia. Since there is no confinement of the waters in the Southern Ocean it is here that most of the deepest bottom water of the Atlantic Ocean originates.

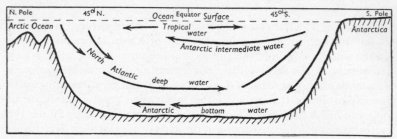

FIG. 253.—Generalized Circulation within the Atlantic Ocean

The fairly dense waters that have been referred to, which sink in the North Atlantic Ocean, do not reach the greatest depths, but form what is known as the North Atlantic deep water, which moves southwards, creeping over the northward-moving bottom water from the Antarctic zone. This North Atlantic deep water permeates the whole of the southern part of the ocean, and finally rises over the denser Antarctic deep water which is descending from the surface in high southerly latitudes (Fig. 253). This Atlantic deep water is accompanied by dense saline water which flows out of the Mediterranean Sea basin as a deep current below an eastward-flowing surface current passing through the Straits of Gibraltar from the Atlantic Ocean. The circulation of the water at the Straits of Gibraltar is of interest, and is shown by Fig. 254. It will be seen that there is a shallow sea over the sill at the Straits of Gibraltar which separates the Atlantic Ocean from the Mediterranean Sea. There is a strong surface current of shallow depth, bringing fairly warm saline water from the west. This current is partly accounted for by the gradient which is produced through the lowering of the surface level of the inner parts of the Mediterranean Sea by great evaporation. This evaporation is accompanied by an increase in salinity, and some water sinks as a dense layer to the floor of the Mediterranean Sea. There is set up a difference of density which produces a return flow of saline deep water into the Atlantic Ocean over the sill at the Straits of Gibraltar and this sinks and joins the Atlantic deep water. It is usual for an inward-moving surface current to be above an outward-moving deep current at an opening from an ocean into a more saline marginal sea. The reverse is true of a marginal sea, which has relatively fresh water in comparison with the outer ocean water. Thus, there is a westward surface current over a deep current moving eastwards in the Skaggerak and Kattegat between the North Sea and the Baltic Sea. Two comparable currents flow between the Mediterranean Sea and the fresher Black Sea.

Hence, there is a stratification of the ocean waters, and the section of Fig. 253 shows several types of water from various sources at different levels. It is certainly quite common for warm but highly saline water to lie below relatively cold water that is only slightly saline. This is exemplified by what is known as Antarctic intermediate water which, although quite cold, is diluted to such an extent that it is not so very dense. It flows northwards in the South Atlantic Ocean above somewhat warmer southward-flowing North Atlantic deep water and below surface tropical waters. There has come into being a conception of 'water masses' in the oceans which are analogous to air masses in the atmosphere, and with movement a 'water mass' may be brought to a position where it mixes with water of a different mass, so that it is transformed as regards its temperature and salinity.

The vertical circulation, which seems typical of the Atlantic Ocean as shown by Fig. 253, is not found in the Pacific or Indian Oceans. The bottom and deep waters of the South Atlantic Ocean apparently spread eastwards and enter the Indian Ocean. Farther east still this water serves the Pacific Ocean. In both of these oceans a return of these deep waters seems to take place with a reversal of the northward flow in the northern parts of the oceans. Just as in the Atlantic Ocean, there is some supply of water to the deep waters of the Indian Ocean from the Red Sea. This supply is somewhat analogous to that which moves from the Mediterranean Sea into the Atlantic Ocean. It is almost certain that much, if not all, the deep water in the Pacific Ocean is of Atlantic origin.

As has been suggested, there must be a compensating elevation of water in certain areas of the oceans to make good the loss to the surface of water which sinks. It has been shown that deep water rises over the denser Antarctic bottom water as it sinks in the South Atlantic region, but there are, of course, several zones of upwelling such as those off the western coasts of Peru and South-west Africa. Quite obviously the whole circulation is extraordinarily complex, but

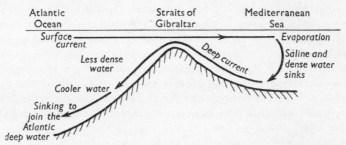

FIG. 254.—Movements of Waters at the Straits of Gibraltar

there is little doubt that the surface circulations are merely one aspect of the complete picture which, as yet, is far from being fully understood.

Effect of Oceanic Circulations on Climate and Weather

Throughout the studies of climatology, in particular of those regions in middle and high latitudes, the contrast between continental and marine areas is frequently presented. Although the land masses undoubtedly play their part in producing this contrast, there is an oceanic influence as well. As Fig. 250 shows, there are westward-moving sea currents near the Equator which bring water to the eastern shores of the land masses, and from here surface water, that is obviously quite warm, moves into high latitudes to become eastward in flow. Within the tropics the contrasts in temperature between maritime areas off the west and east margins of the continents are never very great. However, the coastal fringes on the west of hot deserts are somewhat cooler than most other coastlands in the tropics (see p. 293).

In temperate and high latitudes the general eastward flow of winds and water is of importance. On eastern shores in the winter, cold winds of continental origin lower the temperatures of marine zones, but the effect of relatively warm water offshore is to ameliorate the winter chill of the coastal regions. West coasts of continents in cool temperate latitudes are favoured in winter by the arrival of warm water of tropical origin, and their mean temperatures are often remarkably high for the latitude. This is shown by the poleward bends in the reduced isotherms. On the other hand, west coasts in warm temperate latitudes may show the effect of cold water offshore by the equatorward loops in the isotherms, as, for instance, along the coasts of Northern Chile and South-west Africa. However, the isotherms bend polewards on eastern shores in these latitudes because of warm currents offshore. A comparison of two places on opposite coasts of an ocean in approximately the same high latitude shows that there is usually a greater annual range of temperature on the west side of the ocean than on the eastern side.

A brief reference must be made to one other effect of the oceans. It can be shown that in oceanic areas such as those affected by the Gulf Stream and its related drifts, or by the Kuro Siwo and its extensions, there are large areas of water surface that are relatively quite warm, and from such surfaces there is much evaporation into the cooler overlying air. Furthermore, it may be shown that there is a great transfer of energy from the ocean water to the atmosphere above as a result of this evaporation. It so happens that this access of energy to the atmosphere occurs in the neighbourhood of the average position of the polar front in both the North Atlantic and North Pacific Oceans. Thus there appears to be a close relationship between these currents and drifts and frontal developments, so that it is suggested that the ocean water here is the source of the energy of depressions. Finally, since cyclonic circulations affect winds which in their turn control surface currents to a large extent, it is possible to envisage a complete interrelation between the oceanic and atmospheric circulations which will be more fully understood as more detailed knowledge becomes available.

Tides

The tides in the oceans are a study in themselves, and it is intended to give here a very brief résumé of the processes which are involved.

(a) *Effect of the Attraction of the Moon and the Sun.*—A tide is the periodic elevation and depression of the ocean surface which occurs locally very nearly every $12\frac{1}{2}$ hours. There would be a pair of high tides and a pair of low tides at any one time if the earth had a uniform cover of water. Fig. 255 shows this, and it will be observed that the two high tides (H_1 and H_2) are antipodal in position, so are the two low tides (L_1 and L_2). The explanation of this phenomenon, reduced to the simplest terms, may be followed by reference to Fig. 256. The attraction of the earth and the water of its oceans towards the moon is dependent upon the distances between the moon and H_1, the moon and the centre of the earth, and the moon and H_2. The water at H_1 is attracted more than the earth, and the latter is attracted more than the water at H_2. The fluid nature of water permits a distortion of its surface, and so produces the high tides and the low tides. It may be mentioned that the pull of the sun is also of some importance in raising tides in much the same way as that of the moon; the high tides are less elevated and the low tides are less depressed than those caused by the moon.

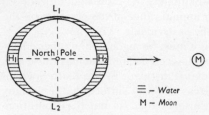

FIG. 255.—Tides produced by the Moon

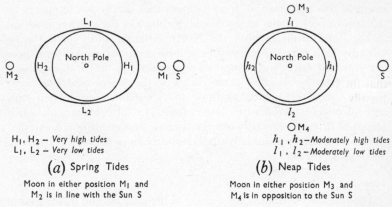

H_1, H_2 — Very high tides
L_1, L_2 — Very low tides

(*a*) Spring Tides

Moon in either position M_1 and
M_2 is in line with the Sun S

h_1, h_2—Moderately high tides
l_1, l_2 — Moderately low tides

(*b*) Neap Tides

Moon in either position M_3 and
M_4 is in opposition to the Sun S

FIG. 256.—Spring and Neap Tides

As the earth rotates on its axis, the high and low tides progress in longitude. Assuming that there is a high tide at a given line of longitude, then a half-rotation of the earth will not bring this line into a position to receive its next high tide (caused by the moon) since the moon has moved in its orbit around the earth during this interval, and this delays the onset of the following high tide by about 26 minutes.

During a lunar month the moon moves completely round the earth, and during that period the sun and the moon are continually affecting the earth as regards its tides. The sun may be regarded as fixed in position, and thus on two occasions in each month the moon and sun collaborate to give very high and very low tides (spring tides) and on two other occasions their influences are in opposition, and they produce less high tides and less low tides than usual (neap tides) as shown by Fig. 256.

(b) *Progression of Tide Waves.*—The theory of a progressive wave of high water moving across the surface of the Southern Ocean (as the earth rotates from west to east this wave would move westwards) and sending branch waves northwards through the other oceans is now considered erroneous.

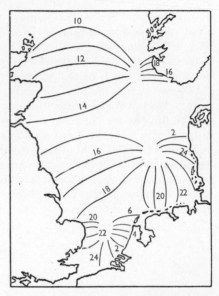

FIG. 257.—Co-tidal Lines in the North Sea
Figures show hours of high tide.

Today tide waves are explained by considering large sections of each ocean as enormous tanks of water, rectangular in shape. The water in such a tank may be made to rock, so that its surface oscillates in a manner which is dependent upon the depth of water and the length of the tank. The tide-producing forces described above set up the oscillations in the form of stationary waves which move across the ocean surface in a succession of high and low tides. Each section of an ocean surface which can be recognized as oscillating in a particular way is determined very largely by the depth of water in that zone. In marginal seas, and in parts of the open ocean, the dimensions often prevent direct waves of this type from being set up, but tides are induced in such areas from neighbouring sea surfaces which are oscillating under the tide-producing forces.

As a result of the effects of the rotation of the earth, the tide waves must be regarded as pulses of high and low water which are gyrating around fairly well-defined points in the seas. For instance, in the North Sea there are three such points, and the radiating lines in Fig. 257 show by the sequential numbers the progress of high water around the coasts.

(c) *Effects of Tides on Shore Regions.*—In the open ocean the difference between high and low tide is usually very small, of the order of two or three feet. In many cases the tide experienced on the shores of continental regions is caused by a progressive wave which has been induced by the pulsations of a stationary wave in the nearby seas. This progressive wave, on entering the shallow water over the continental shelf, tends to increase in height, and give tidal ranges of several feet and in some cases as much as forty feet.

In bays and estuaries the effects are quite varied. On entering an estuary the tide is rising against the gradient of the river-bed and the flow of the river. The rising tide may, in narrowing estuaries, produce a wall of water three or four feet in height, as in the case of the Severn bore with spring tides. With the fall of the tide in an estuary there is the additional force of the river flow which has been temporarily dammed, and this is helpful in creating effective scour.

In bays, such as the Bay of Fundy, the water apparently oscillates as an independent unit with a stationary wave. The exceptionally high tides of this bay are caused by its decreasing width, which produces a range of fifty feet at its head on some occasions.

The double tides of Southampton have been explained in various ways, but most explanations are erroneous and none is completely satisfactory. The most recent view is that the phenomenon is the result of combined tidal pulses, but the exact nature of this combination is as yet unknown.

INDEX

(References in heavy type indicate illustrated items.)

A

Abandoned meander course, **88**, 110
Ablation (snow), 94
Abrasion, 8, 119
Absorption of heat, 163
Acacias, 243, 279, 306, 323
Accumulation (snow), 94
Acidity (soil), 260
Adaptation (plants), 240, 244, 251
Adiabatic changes, 189–97, 224
— lapse rates, 189
Adret, 163
Advection fog, 205, 214, 328
Aeolian deposition, 118–21
Ages of mountain building, 51 *et seq.*
Aggradation, 62
Air masses, 206–9
— source regions, 206, **208**
— stable, 206–7
— types, 206
— unstable, 206–7
Air pressure, 171 *et seq.*
Aletsch glacier, **95**, 108
Alkalinity (soil), 260, 330
Alluvial fans (cones), 122, **123**
Alluvium (soil), 263
Alpine glaciers, 93, 97
Alpine-Himalayan folds, 36, 54
Alps, 36, 39, **40**, **41**
Alto-cumulus cloud, 199, **200**
Alto-stratus cloud, 199, **200**, 213
Anabatic winds, 223
Anaerobic soils, 257
Anemometer, 177
Anomalies (temperature), 169
Antarctic beech, 323
— water, **352**
Antarctica, 51, 93
Antecedent drainage, 90
Anthracite, 18
Anticyclones, 218
Anticyclonic gloom, 219
— gyrals, 346, 349

Appalachian Mountains, 32–7, 51,
128
— Valley, **37**, 79
Aquifers, 17
Arctic glaciers, 93, 98
— Ocean, 337 *et seq.*, 340
Arêtes, 99, 102
Arroyos, 123
Artesian water, 17
Artificial rain, 203
Atlantic coasts, 144–5
— Ocean, 337 *et seq.*
Atolls, 152–4
Azonal soils, 263
Azores, 302, 319, 339

B

Backwash, 138
Bacteria (soil), 255–7
'Bad lands', 121, 126, 265, 321
Bajada, 123
Baobab, 243, 279, **288**
Bar (pressure), 172
Barchans, **120**
Barogram, 173
Barograph, 172
Barometer, aneroid, 172
— mercury, **171**
Barrier reefs, 152
Basalt, 13, 24
Base level, **62**, 86
Basin and range areas, **46**
Basins (soil conservation), 265
Bathylith, **23**, 35, **40**
Bathymetric contours, 338
Bauxite, 259, 294
Beach deposits (lakes), 108
Beaded esker, **115**
Beaufort Scale, 177
Benthic organisms, 340
Berg wind, 223, 302
Bergschrund, **98**, 100, 101
Berne (site), 89

359